Paolo Bondavalli
2D Materials

Also of interest

Crystallography in Materials Science.
From Structure-Property Relationships to Engineering
Schorr, Weidenthaler, 2021
ISBN 978-3-11-067485-9, e-ISBN (PDF) 978-3-11-067491-0

Active Materials
Fratzl, Friedman, Krauthausen, Schäffner, 2021
ISBN 978-3-11-056181-4, e-ISBN (PDF) 978-3-11-056206-4
⌾ Open Access.

Magnesium Materials.
From Mountain Bikes to Degradable Bone Grafts
Oshida, 2021
ISBN 978-3-11-067692-1, e-ISBN (PDF) 978-3-11-067694-5

Nanostructured Materials.
Applications, Synthesis and In-Situ Characterization
Terraschke (Ed.), 2023
ISBN 978-3-11-045829-9, e-ISBN (PDF) 978-3-11-045909-8

Chemistry of Carbon Nanostructures.
Müllen, Feng (Eds), 2017
ISBN 978-3-11-028450-8, e-ISBN (PDF) 978-3-11-028464-5

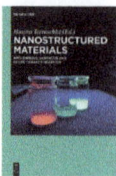

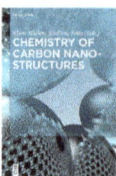

Paolo Bondavalli

2D Materials

And Their Exotic Properties

DE GRUYTER

Author
Dr. Paolo Bondavalli
Thales and Research and Technology
110 résidence des eaux vives
91120 Palaiseau
France
paolo.bondavalli@thalesgroup.com

ISBN 978-3-11-065632-9
e-ISBN (PDF) 978-3-11-065633-6
e-ISBN (EPUB) 978-3-11-065638-1

Library of Congress Control Number: 2022932185

Bibliographic information published by the Deutsche Nationalbibliothek
The Deutsche Nationalbibliothek lists this publication in the Deutsche Nationalbibliografie;
detailed bibliographic data are available on the Internet at http://dnb.dnb.de.

© 2022 Walter de Gruyter GmbH, Berlin/Boston
Cover image: Gettyimages/LuckyStep48
Typesetting: Integra Software Services Pvt. Ltd.
Printing and binding: CPI books GmbH, Leck

www.degruyter.com

To my wife (for her brightness and patience), to my soccer-girl Camille (the best right-defender of the next generation) and to all my friends who were supportive during this adventure and help me each day to advance (in alphabetic order and not in order of importance!): Costas (my great Greek friend), Gaetan (for all the discussions around the spray-gun machine), Gilles (for his true friendship and for time spent to do fitness), Louiza (for all the fruitful discussions), Pierre (and his way of tackling life), and finally to my mother that pushed long time ago to do physics at school (I did not want!!!). Thank you to all of you!

Foreword

After the advent of graphene in 2004, celebrated by the 2010 Nobel Prize in Physics awarded to Andre Geim and Konstantin Novoselov for "groundbreaking experiments regarding the two-dimensional material graphene", there has been a surge of related articles on 2D materials, either peeled off from lamellar crystals existing in nature – by analogy with the exfoliation of graphite – like the transition metal dichalcogenides or black phophorous, or, instead, upon synthesis of novel artificial ones, especially the so-called Xenes, following the first realization of silicene in 2012.

In 2022, a number of books in this area and a plethora of review articles surfing on the wave have been already published. On the one hand, the field is still in full expansion; on the other hand, it is nurtured by fantastic discoveries, great expectations and a strong hype.

To be fair, I have to say that many books are quite similar in content, and the review articles, which can be useful, unfortunately often look like uninspiring bailiff reports.

In such a context, why preface a new book when the series is already overabundant?

The reason is a stark contrast. Unlike most cases, this book is structured in a very different way, while its author, Dr Paolo Bondavalli, gives his own compelling opinion and far-reaching personal views.

The atypical journey of Dr Bondavalli, who went to a leading institute to do fundamental research work after spending a decade in a company, likely explains the book's peculiar architecture and its ambitious goals.

The reader will enter the realm of 2D materials and understand the fundamentals of the topology involved without tedious mathematical developments. He will discover fascinating exotic properties and target groundbreaking potential applications in the "more than Moore perspective" – notably for thermoelectricity – and at the "beyond CMOS horizon", for example for all-spin logic.

He will learn how materials such as stanene, few-layer black phosphorous or twisted van der Waals layers are arranged in a Lego architecture, along with the implementation of straintronics, and revolutionary paradigms such as valleytronics or twistronics could lead to radically new low-energy consumption devices for an environmentally friendly sustainable world.

Finally, this book will galvanize the interest in 2D materials, help overcome inherent hurdles, provide avenues for the fabrication of reproducible innovative devices with long-term stability in ambient conditions and pave the way for their production at large scale, provided, of course, that the manufacturing costs will be low enough.

Guy Le Lay, Aix-Marseille University
Marseille, January 17, 2022

https://doi.org/10.1515/9783110656336-202

Author's premises

This book has been a real challenge that I wanted to tackle. It took me $2\frac{1}{2}$ years to achieve it. I was very curious to explore the exotic properties that are emerging in 2D materials by using technology such as straintronics. Everything started some years ago. I have been working in a large company as senior researcher for many years, maybe too many. I was going too far from all the things that motivated me to spend my life trying to do science. I wanted to come back to physics and to explore new "things". "Things" and phenomena that could make me dream yet. I wanted to find a thread that it will be able to give me new motivation or at less to find the spirit that pushed me to make research 20 years ago. I have written a previous book on graphene properties but this one is completely different for me. I have been working on graphene for more than 10 years now. Graphene and its applications are in my so-called comfort zone. I needed something more complex that allows me to go beyond my knowledge challenging my conception of physics. I was pushed to focus my attention on exotic properties of materials, firstly by working on topology matter. Indeed, I think we are on the verge of a real revolution that will change completely our way of conceiving physics and so devices. This revolution will have a terrific impact in our everyday life. Two-dimensional topological insulators are extremely interesting objects of research that are disclosing their incredible potential in terms of physics behind and its applications. This revolution started some years ago with the discovery of quantum Hall effect that opens new research horizons. However, the potential impact of materials featuring the quantum spin Hall effect can change radically the present technology and push finally us in the "beyond CMOS realm". I have always thought that physicists have to dream and make people dream. We do not have to continue to do the same things simply changing the materials as a function of the different science fashions. We have to think out of the box conceiving a new way of processing information exploiting devices that cannot be labelled precisely using the present technology. Indeed, more than 100 years ago, the "transistor" was not a concept. Now our technology depends completely on this component and each system is thought as a function of it. Processing charge is a sort of postulate of the modern technology. For this reason, when we discover new materials, as in the case of 2D some years ago, the first thing that scientists did was to fabricate transistors with them. In my opinion, 2D material transistors are really a misguiding direction. Two-dimensional materials will help us enter finally in a new domain of science and new concepts of devices acting differently and not labelled in the same way. The definition of sensors, transistors or memories will not have any meaning in the future considering that new devices and systems will be able to process and transport information at the same time.

In this book, I have identified some topics. Each chapter is motivated by my will, and my curiosity, to understand the physics of each exotic phenomenon. Each chapter by itself is a challenge. Because of that, I tried to write it using simple terms

https://doi.org/10.1515/9783110656336-203

and to explain the great lines of each one identifying the main physics behind and the potential implementation in the real world. I tried to avoid too complex mathematical approaches because it was not the aim of this book to give an exhaustive description of everything. I hope that the reader will enjoy this book that pushed me in new unknown "lands" where I did not lose myself, but inversely, I was able to find again the motivations to do physics and to believe in science yet.

I want to dedicate this book to all the scientists, also around me, who think that physics has to be respected. To all the people who when they speak about physics they know what they are saying.

Thank you for your attention and enjoy!

Contents

Introduction

This book is conceived for researchers in different fields of physics or engineering or chemistry who want to have a global overview of the perspectives given by the discovery of exotic properties in 2D materials. The book's aim is to give some hints and to make understand easily the physics behind. To have a deeper understanding of each chapter, the reader will be able to follow the references quoted in the book.

In the first chapter, we deal with topological insulators. We introduce some concepts such as the topological invariant and we talk about the difference between the quantum Hall effect and quantum spin Hall effect. We present stanene and its properties that are extremely interesting, considering that it is a material showing topological features at ambient temperature. Finally, we discuss about the application in the field of thermoelectricity, where 2D topological insulators could be a real turning point.

In the second chapter, we explain what the magic angle is and present the pioneering contributions in the field, also using an historical approach of the discovery. After a short introduction on superconductivity, we will explain how, starting from the theoretical predictions, scientists were able to demonstrate superconductivity in bilayer graphene.

In the third chapter, we show the main features associated with valleytronics and why this last could constitute a revolution in the field of high-density information storage systems implementing optically driven supercomputers. We point out how for each charge we are able to store three main data related to charge, spin and valleys. We will also deal with quite short-term applications such as quantum distributed key that will be able to start a new revolution in cryptotechnologies, thanks to the intrinsic valleytronic properties of transition metal dichalcogenides such as MoS_2.

In the fourth chapter, we deal with black phosphorous, from its discovery to the renewed interest after the 2D revolution and the graphene "discovery". We discuss its main physical characteristics and its potential implementation in real devices (e.g. optoelectronics, energy storage and solar cells). Moreover, we introduce the main passivation strategies that allow handling this material extremely reactive in air or in contact with humidity/water.

The fifth chapter is devoted to straintronics and how this technique can be used in single 2D layers or in van der Waals 2D architectures to generate ad hoc multifunctional structures. We perform a state of the art of the main example of application and present some perspectives.

Finally, in the last chapter we perform the roadmap for each topic and we present our personal vision. This vision is based on the global overview that we have acquired

https://doi.org/10.1515/9783110656336-001

in the field and on the bottlenecks (technological, cost, fabrication, etc.) to overcome to fabricate devices exploiting the properties described. Moreover, we evaluate the potential of innovations and if they are more related to a "more Moore", "more than Moore" or "beyond CMOS" approach.

1 2D topological insulators and quantum spin Hall states: what are they and which are the potential applications for the real world?

1.1 Introduction

Two-dimensional (2D) topological insulators (TIs) are a new class of materials that can, thanks to their extraordinary properties, finally lead us to the beyond Complementary metal oxide semiconductor (CMOS) world. In this first chapter, we will explain in a simple way, what is the meaning of TIs, the physics behind them and we will try to identify the main breakthroughs that can be brought by their implementation in the real world. We will mainly focus our analysis on the potential of TIs for thermoelectric (TE) considering the huge potential breakthroughs in this field. However, firstly, we would like to explain the signification behind the name and especially the word "topological". Topology can be defined as a sort of modern version of geometry. Indeed if we consider the definition of the famous German mathematician Felix Klein [1], what distinguishes different kinds of geometry is the kind of transformations allowed until we realize that something has changed. If we consider the Euclidean geometry, in this case, it is clear that we can apply different kinds of transformations (e.g. symmetry as a function of a line or as a function of a point). However, we cannot stretch the geometrical shape. The geometric shapes (e.g. triangular and square), after the Euclidean transformation, can lay one on the other with a perfect match. We called it "congruence". In topology, any continuous change that is completely reversible is allowed. For this reason, for example, starting from a circle we can obtain only a triangle only because changing the position of the points we can change the shape. Briefly, we can say that topology explains how an object shape can be completely deformed into new one without losing its core properties. To give a quite short definition of the term "continuous", we can say that it means that we can move without changing the nearby points position of each other. One of the most known examples is the mug and the doughnut case. Indeed, it is possible to modify the shape of a doughnut and make it become a mug respecting the topology of the object. The hole in the centre of the doughnut becomes the ear of the mug in a logic way (see Fig. 1.1).

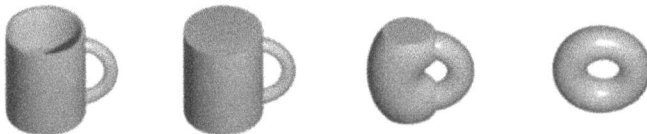

Fig. 1.1: How a mug can be transformed in a doughnut (which is a topologically equivalent object) exploiting homeomorphism (also known as topology homeomorphism).

https://doi.org/10.1515/9783110656336-002

The main conclusion of this transformation is that the two objects are topologically equivalent. For this reason, topology can be familiarly referred to as "rubber sheet geometry". In opposite, we can also observe that in case of a sphere and of a doughnut, we have two objects that are not topologically equivalent. Each one is distinguished by a number called *genus* which simply corresponds to the number of holes in the shape. This number is an integer number and it corresponds to the Chern number (we will see this concept in a next section in this chapter) in topology. In case of two different *genus* numbers, the surfaces cannot be considered topologically equivalent. The question now is *how can we define a topological invariant that characterizes a specific object with a specific surface?* For surface, as mentioned by C.L. Kane [2], we can use the theorem of Gauss–Bonnet [3], which states that the integral of the Gaussian curvature K over a surface defines an integer topological number, called the Euler characteristic [4]. If we analyse one of the easiest cases, a sphere, we obtain

$$\chi = \frac{1}{2\pi} \int_S K dA \tag{1.1}$$

In this case, $K = 1/R^2$ (where R is the radius of the sphere), and so the Euler characteristic is 2.

(a)

(b)

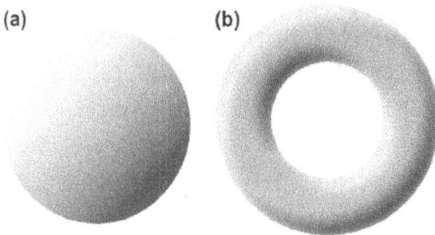

Fig. 1.2: Topologically non-equivalent objects: (a) a sphere and (b) a torus (a doughnut).

In general, considering the *genus* number, the Euler characteristic is obtained through the following equation: $\chi = 2-2g$. If we come back to the doughnut (called also a "torus", scientifically speaking), we obtain in this case 0 considering that we have a single hole (see Fig. 1.2). Indeed, this can also be explained by the fact that in a torus the external points have a positive curvature and the internal ones a negative. Therefore, they cancel each other out. The logic conclusion is that the two objects are not topologically equivalent. Taking into account all the previous examples and explanations, we start having the keys to understand why we can use the term "topological". Now we have to transpose it to materials and more specifically to nanomaterials. What we have to keep in mind is that the main feature of topological materials, considering what already told, remains robust regardless of the deforming forces applied. In this context, in the next section we will analyse specifically the meaning behind the words "topological insulators".

1.2 But indeed, in simple terms, what is a topological insulator (using quantum mechanics)?

We have explained in general terms the meaning of topology. Now, we have to move forward and to understand better why we can use the topological classification also for matter. But now it is time to talk about insulators. The common definition of an insulator is a material that owns a gap. This gap is an energetic separation between the excited states from the ground state (usually called conductive band (CB) and valence band (VB) for semiconductors). As outlined by Kane [2], we can say that two insulators are topologically equivalent if we can move from the band structure (in case of the analogy, the shape) of one to the other and this making smoothly evolving the Hamiltonian in order to keep the system in its ground state (the genus number does not change, i.e. the number of holes). This can be achieved, thanks to an adiabatic continuity between the two insulators, where an adiabatic process is commonly defined as a process where there is no transfer of heat/mass of substances between a thermodynamic system and its environment. If we put this process in the context of quantum mechanics, we observe that a system governed by a Hamiltonian with time-dependent parameters can evolve from an initial eigenstate to an eigenstate at later times when the parameters change in an extremely smooth way as a function of time. To be more specific, the word "eigenstate" comes from the German/Dutch word "eigen", meaning "inherent" or "characteristic". An eigenstate in quantum mechanics consists in the measured state of some objects possessing quantifiable characteristics such as position and momentum. Briefly, if in case of two-band insulators, a scalar function can be defined in the momentum space characterizing the wave function overlap between Bloch states in the two insulators. Two insulators can be defined as adiabatically linked, when the superposition of the wave functions is nonzero for all momentum in the Brillouin zone. The main consequence is that we can deform one insulator (or to be more precise, to move from the state of one insulator to the state of the other one, where for state means all the parameters describing the geometrical, energetical, and other states) into the other gradually without closing the band gap, like in case of a topological transformation of two "topologically equivalent" objects. We also observe that the adiabatic path keeps intact all the symmetries of the insulators [5]. In the very interesting paper of Gu and co-workers [5], it is well explained that in the other case we have a so-called quantum phase transition. If we analyse the situation in the optics of the adiabatic continuity, we discern a quantum phase transition if we do not have an adiabatic path between the quantum systems' ground states. In this case, it is impossible to modify adiabatically one quantum system into the other, without experiencing some intermediate phases (which are not insulating ones). These two different quantum states are relevant of two different quantum phases of matter. The point where the phase changes and that takes place when we modify one state in another is commonly called a quantum phase transition point. When we talk about the quantum phase transitions, we can identify two main ones: Landau type and topological ones, depending on the origin of the phase transition point. In the first case, the two quantum phases separated by a quantum phase transition do not own the same symmetries: the two quantum states are not able to mutate progressively into each other without passing through a quantum phase transition. In the second case, the two quantum phases have exactly the same symmetry; however, their ground state wave functions do not feature the same topological structure. When a finite energy gap exists between the ground state and the excited ones (we commonly described this situation with a gapped energy diagram), the topology of the ground state wave function is not allowed to change through any adiabatic process without closing the energy gap. Consequently, if the ground state wave functions of two gapped quantum systems have different topologies, as we try to modify one into the other, a phase transition point appears, at which the energy gap closes and the ground state wave function transforms its topology. This topological transition can happen even without interactions, for example, in non-interacting band insulators.

1.3 How to move from the quantum Hall effect to the quantum spin Hall effect? (Intuitive explanation without mathematical developments)

As explained by Qi et al. in 2010 [6], the different states of matter such as magnetism, superconductors and crystalline solids are often classified as a function of the symmetry that they spontaneously break. The first exception to this rule is constituted by the quantum Hall effect (QHE). This effect was discovered in 1980, thanks to the first experiments exploring the quantum regime of the Hall effect that were performed in 1980 by von Klitzing, using samples prepared by Dorda and Pepper [7]. What is hidden behind QHE? This phenomenon takes place when a magnetic field is applied to a 2D gas of electrons in a semiconductor. Briefly, a 2D electron gas is a quite simple model in solid-state physics where the electrons are free to move in 2D, but tightly confined in the 3D. When we apply a low temperature and a high magnetic field, these states will be localized at the edges of the semiconductor. Finally, what we obtain is two counter-flows of electrons of the semiconductor that are spatially separated and are localized at the top and bottom edges of the semiconductor. Now, we will try to explain in simple terms what happens. In 2D, when classical electrons are subjected to a magnetic field, they follow circular cyclotron orbits. Indeed, the main effect of the magnetic field, and low temperature, is that the electrons will follow small circular paths with a radius, at the edges of the sample, which is inversely proportional to the magnetic field intensity with quantized degenerate quantum states where the degeneracy depends on the magnetic field strength (in this case, directly proportional to B). Therefore, when the system is treated quantum mechanically, these orbits are quantized. The energy levels of these quantized orbitals acquire discrete values. These orbitals are known as Landau levels. For strong magnetic fields, each Landau level is highly degenerate and therefore a high quantity of electrons has exactly the same energy. Why? Because increasing the magnetic field to very high levels (>10 T), we reduce the number of states to only some highly degenerated. For sufficiently strong magnetic fields, each Landau level may have so many states that all of the free electrons in the system sit in only a few Landau levels and this is the precise reason why in this regime we observe the QHE. These states are robust in the sense that the electrons can propagate in the edge without dissipation, and the electrons, when they encounter an impurity, keep going avoiding it. The main drawback associated with this effect, with its realization, is that we need to apply a strong magnetic field and to reach very low temperatures. It is important to understand correctly the QHE in order to grab the signification of the quantum spin Hall effect (QSHE), which is familiarly speaking, its "cousin". We can say in an intuitive way that the QSHE is the correspondent of the QHE in case of spinful chain (we take into account the spin of the electron in the edges). The meaning is that in case of QSHE we have two lanes which are localized at the edges of the semiconductor: one having forward spin-up and backward spin-down moving electrons, and one having forward spin-down and backward spin-up moving electrons. These two lanes can be split without any external magnetic field applied, a clear advantage compared to the QHE (also in terms of energy consumption). A system with these characteristics is said to be in a Quantum spin hall (QSH) state considering that we observe a net transport of spin forward in the top edge and backward in the bottom edge exactly in the way it happens for QHE. A system in a QSH state is a TI. The word "topological" has its proper signification considering that the states in the top and bottom edges are robust ("topologically protected"). To mention the reference work of Qi and Zhang in order to understand in an intuitive way the meaning of topologically protected, please take into account an analogy between what happens in case of a surface with an antireflection layer.

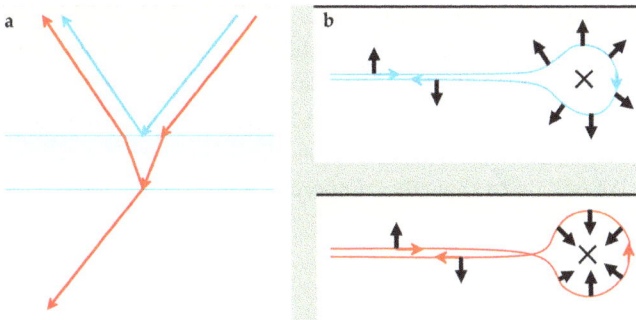

Fig. 1.3: Analogy between the destructive interference phenomenon and the backscattering at the edge of a topological robust structure (reproduced from [6], with permission of the American Institute of Physics).

In this case, the interference of two light waves reflected is negative; hence, there is no reflection (see Fig.1.3b). What happens to a QSH state is very similar. If we consider an impurity, in this case, an electron moving from left to right with spin-up can move in two ways around an impurity: clockwise or anticlockwise. The two paths differ by a full 2π rotation of the electron spin which leads to a destructive interference between the wave functions of the particles and so of the paths and therefore to a perfect transmission. However, if the impurity is magnetic, the time reversal (TR) symmetry is broken and the reflected waves do not interfere destructively anymore. A simple question can be raised: what is the TR symmetry? We can say that a physical system is TR invariant if its underlying laws are not sensitive to the direction of time. To explain in a simple way, in our case the configuration that we are analysing, the moving of electrons in lanes is completely topologically protected by the fact that this configuration is time invariant: the configuration for a time t is completely equivalent to a configuration at a time $-t$ and so backscattering is not possible (see Fig. 1.3). A magnetic impurity could theoretically break this symmetry. The materials showing the QSHE are also called TIs because they usually show metallic behaviour in the edges (surface in 3D and borders in 2D materials) and insulating behaviour in the bulk. To explain more specifically why we use the terms of TIs for semiconductors showing QSH states, we have to take into consideration that the QSH states can be identified only in case of single pairs. We will try to provide some examples in the next paragraphs.

1.4 Topological matter and topological invariant

In this section, we will take inspiration in part from the pioneering work of Kane [2] and specifically from his book on TIs. We will try to explain in the simplest way some of the most fundamental concepts of topology. In some parts, it will be difficult to grab the physical meaning of them in an easy way, considering that we will deal with concepts quite distant from our normal perception of things. We will avoid too complicated mathematical developments, considering that these last can be easily found in references (quoted in the book). We will focus our attention, as in the whole book, on the physics behind adopting an intuitive perception of the different phenomena presented. Chern number and Z_2 invariant are used to identify and define Chern insulators and non-trivial TIs. We will try to show their importance and significance.

1.4.1 What is the Chern number?

In this section, we will be able to describe the ground state of systems enduring a magnetic field B, using a topological invariant which is called the Chern number. This is particularly important in case of QHE, as explained before. Before trying to explain the physical meaning of the Chern number, we have to define the Berry phase [8], another fundamental concept in topology. Indeed, Berry phase has an important role in topological matter physics. The Berry phase emerges when we consider the intrinsic phase ambiguity of a quantum mechanical wave function. This is not what we can call an intuitive explanation. We can say, in simple terms, that the Berry phase in classical and quantum mechanics is a phase difference gained over the course of a cycle, when a system adiabatically gets through cyclic processes, strictly resulting from the geometrical properties of the parameter space of the Hamiltonian. We will now try to obtain with some simple mathematical passages an equation describing the Berry phase and its impact on Chern number definition. If we consider a quantum system at the nth eigenstate, an adiabatic progression of the Hamiltonian means that the system does not move from the nth eigenstate; however, it gains a phase factor. There are two main contributions to this phase that can be attributed to the state's time evolution and another due to the change of the eigenstate according to the progression of the Hamiltonian. The second term is defined as the Berry phase for non-cyclical modifications of the Hamiltonian. This term can vanish if we adopt a different choice of the phase associated with the eigenstates of the Hamiltonian at each point during its progression (between instant 0 and t). However, if the variation is cyclical, the Berry phase will not vanish and becomes invariant and an observable property of the system. If now we consider the proof of the adiabatic theorem given by Max Born and Vladimir Fock [9], we can point out the entire variation of the adiabatic process into a phase term. Thanks to the adiabatic approximation, the coefficient of the nth eigenstate under adiabatic process is given by the following equation:

$$C_n(t) = C_n(0)\exp\left[-\int_0^t \langle \varphi_n(t')|\dot{\varphi}_n(t') \rangle dt'\right] = C_n(0)e^{i\gamma_n(t)} \tag{1.2}$$

γ is known as the Berry's phase with respect to parameter t. As previously stated, the Berry phase details the phase gained through an adiabatic cycle. In the present context, it will be useful for classifying loops in momentum space. Changing the variable t into generalized parameters, we could rewrite the Berry phase for any closed loop C in k space and we may define the Berry phase as described in the following equation:

$$\gamma_C = \int_S F d^2K = \oint_c A dK \tag{1.3}$$

where $F = \nabla \times A$ defines the Berry's curvature. Kane's et al. made easier its formalization, suggesting that k is 2D. To be more precise A, called the Berry connection, can be written as

$$A = -i\langle \varphi(k)|\nabla k|\varphi(k) \rangle \tag{1.3a}$$

The generalization to higher dimensions can be easily implemented. The Berry phase can have many physical meanings considering that it details the phase acquired under an adiabatic cycle. In the present context, it will be useful for identifying and detailing loops in the momentum space. Now we consider the simple case of a 2D system.

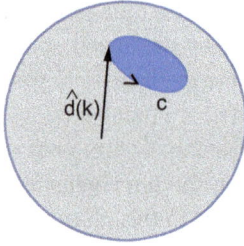

Fig. 1.4: The Berry phase in a two-band theory is obtained by sweeping out half the solid angle through $d(k)$.

As suggested by Kane, it is useful to understand the Berry phase for the simplest two-level Hamiltonian, which may be expressed in terms of Pauli matrices:

$$H = d(k)^* \; \vec{\sigma} = \begin{pmatrix} -dz & dx - idy \\ dx + idy & dz \end{pmatrix} \tag{1.4}$$

This Hamiltonian has eigenvalues $\pm|d|$. A classic result, shown by Berry [10], is that for a loop C in the phase associated with the ground state is, considering Fig. 1.4, given by

$$\gamma_C = 1/2 \; (\text{angle swept out by } \mathbf{d(k)}) \tag{1.5}$$

In particular, when C leads to a 2π rotation of $\mathbf{d}$ in a plane, the Berry phase is π. Indeed, if the Berry curvature is integrated over a closed 2D space (such as a 2D Brillouin zone), then this last is a multiple of 2π that corresponds to the number of times $\mathbf{d(k)}$ vector that surrounds completely the sphere as a function of k. This allows defining the value of a topological invariant defined as the Chern number [11], which for a closed surface S is obtained through the following equation:

$$n = \frac{1}{2\pi} \int_S F d^2 k \tag{1.6}$$

The Chern number quantization has a significance which is not strictly linked only to the two-band model previously presented. Indeed, when we consider a loop C on a surrounded surface, the "inside" of C is arbitrary defined. Therefore, the surface integral calculated over the inside and the outside is logically a multiple of 2π. Consequently, when we integrate the Berry curvature over the whole surface, we obtain that it is logically equal to $2\pi \times n$. This quantization is also strictly linked to the quantization of the Dirac magnetic monopole. Therefore, F is considered as a curvature, using an analogy with the Gaussian curvature K. The Chern number is a concept, which is particularly interesting and useful, in the context of the QHE. Indeed, we know that the integer QHE occurs when a 2D electron gas circular orbit gives origin to the quantized Landau levels. It has been demonstrated that when n Landau levels are filled, keeping the other empty, an energy gap divides the occupied and empty states as exactly what happens for an insulator. However, in a different way compared to an insulator, an electric field leads to the drift of the cyclotron orbits, carrying out a Hall current featuring a quantized Hall conductivity, $\sigma_{xy} = ne^2/h$. We can state that, from this point of view, Landau levels emerge to a band structure. Because the generators of translations do not commute in a magnetic field, electronic states cannot be identified using momentum. However, if a unit cell with an area hc/eB encompassing a quantum flux is addressed, in this case, the lattice translations can commute. Consequently, Bloch's theorem allows identifying states exploiting k, the 2D crystal momentum. Without a periodic potential, the energy levels can be directly identified as the k-independent Landau levels. In an opposite way, with a periodic potential having the same lattice periodicity, the energy levels will tend to dissipate with k heading to the creation of a band structure that appears to be exactly the same of that characterizing an ordinary insulator. The main difference is associated with the topology of the quantum Hall (QH) states compared to ordinary insulators. In a pioneering 1982's paper [12], Thouless, Kohmoto, Nightingale and den Nijs demonstrated that the Chern number is exactly the integer-quantized Hall conductivity. Their demonstration exploited the direct application of linear response theory showing that the Kubo formula [13, 14] for σ_{xy} is exactly identical to eq. (1.6). Starting from intuitive explanation given in the Kane's book, we will show briefly the main hints provided by Laughlin's argument for the integer QHE [15]. In this context, Laughlin schematized an integer QH state on a cylinder where the magnetic flux is raised in an adiabatic way. In this way, the flux passing through the cylinder changes its value from 0 to the flux quantum $\varphi_0 = h/e$.

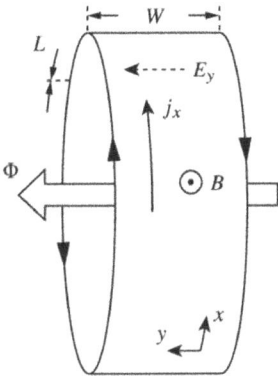

Fig. 1.5: Schematization of the Laughlin's argument exploiting a cylinder. We can observe that electrons are confined on the cylindrical surface when a magnetic field B is applied perpendicular to the surface. A magnetic flux Φ passes through the hole of the cylinder. Consequently, an electric field E_y is applied (reprinted figure with permission from [16], Copyright (2017) by the American Physical Society).

The flux variation makes a Faraday electric field arise, surrounding the cylinder and that is a function of dt through $d\phi/dt$, giving origin to a Hall current which is described by $I = \sigma_{xy} d\phi/dt$ (see Fig. 1.5). Consequently, a net charge $\sigma_{xy} h/e$ moves between the two extremities of the cylinder. We observe that when $\varphi = \varphi_0$, the vector potential is removed by a gauge transformation; therefore, the Hamiltonian come back to its initial state at $t = 0$. Consequently, the quantity of charge transferred is an integer number of electrons $Q = ne$, which directly leads to the quantization $\sigma_{xy} = ne^2/h$. Assimilating this example to a 1D system, the cylinder with the magnetic field applied is a so-called Thouless charge pump with $t = \phi$. The Chern number, which is a pump characteristic, can be obtained by summing over adding all the contributions from all the occupied 1D cylinder subbands of the cylinder. This can be calculated using eq. (1.7), where the Hamiltonian is a function of k and t. Thanks to this equation, we are able to classify the cyclic families of 1D insulators defined by $H(k,t)$ by the Chern number and obtain the topological invariant that characterizes the quantized charge pumped:

$$n = \frac{1}{2\pi} \int_{T^2} F dk\, dt \tag{1.7}$$

But, indeed, what happened in 2D materials that can be related to the Laughlin's argument? We observe that a fundamental consequence of the topological classification of gapped band structures is the presence of gapless conducting states at the edges, where the topological invariant changes. It is as compared to the Laughlin's cylinder to 2D materials where the extremities are the limit between the two domains in the cylinder that corresponds to the interface between 2D materials and the vacuum (the edges). Indeed, in a very intuitive way, they may be understood in terms of the semiclassical skipping orbits of electrons because their cyclotron orbits extend off the edges. The main feature of these electronic states is that they can progress in the edges only in one direction and so they are defined as chiral (see Fig. 1.6).

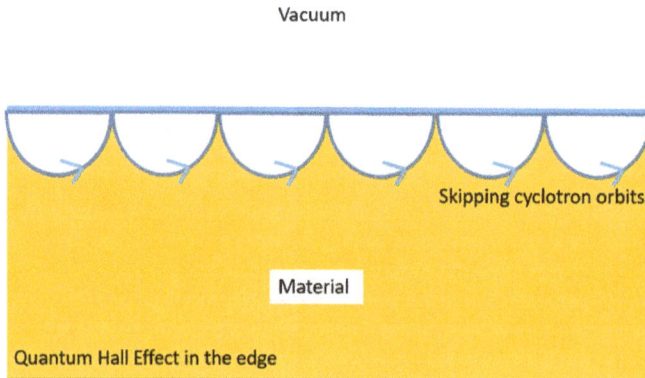

Fig. 1.6: Edge states as skipping cyclotron orbits.

These edge states are robust against any disorder because backscattering is not allowed (without the presence of states implementing it). This points out the perfect quantized electronics transport featured by the QHE. The presence of these peculiar edges' electronics states moving only in one direction is logically strictly due to the topology of the bulk QH state. Consequently, using this analogy, the conductivity in the edge is quantized using the Chern number in 2D materials outlining the QHE.

1.4.2 Z_2 invariant

In the previous sections, we have explained the physical meaning of the Chern number which describes the integer QHE in terms of Laughlin's argument and of a 1D Thouless charge pump. We tried to make intuitively understand why QHE and Chern number are related. We showed that the Chern number can point out the change in the electric polarization when flux $\phi = \varphi_0$ is adiabatically threaded through the cylinder of the Laughlin's argument (which is a very simple and intuitive schematization). In order to define an equivalent formulation for QSHE, we have to transform the cylinder into a Corbino disk (see Fig. 1.7) with a small hole tied by flux. Therefore, in this case, Laughlin's argument details the binding of electric charge to the flux threading the hole. It is clear that this formulation is more general than the non-interacting electron framework that we have been using.

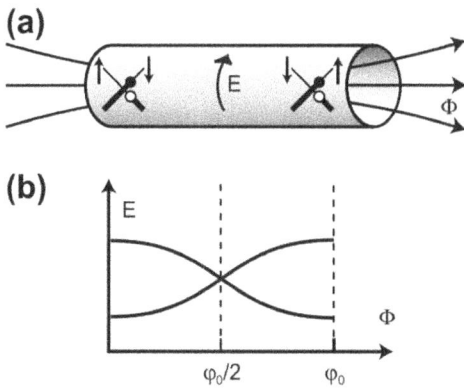

Fig. 1.7: (a) When flux $\varphi_0/2$ passes through a cylinder, the "time-reversal polarization changes". (b) Evolution of the many body energy levels as a function of flux. We observe that when $= \varphi_0/2$ we have a Kramers degenerate state (reprinted from [2], Copyright (2013), with permission from Elsevier).

The Laughlin's argument previously presented is also used to detail an interacting system, and this because the polarization change can be set up in a many body setting. The Z_2 invariant physical significance can be similarly grabbed. Firstly, we have to take into account the same cylinder of the previous schematization, with a specific radius (see Fig. 1.7), so that the eigenstates related to the extremities are quantized (as in the case of QHE and Chern number). From the point of view of a 1D system, we want to verify if there is a Kramers degeneracy in the ground state. This degeneracy is due to all the energy levels that are doubly degenerate in a TR-invariant electronic system if we have an odd number of electrons. When the TR symmetry is verified, the existence of a Kramers degeneracy is simply linked to the number of local electrons (even or odd). The Laughlin argument, already used for Chern number and QHE, can also be applied to an interacting system because the polarization change is well addressed in a many-body setting. As pointed out by Kane, we can refer to this Z_2 quantity as the "TR polarization," or the "local fermion parity" (for more details, read the Kane's book [2]). Like the polarization in Laughlin's argument, the TR polarization is also identified in an interacting system. The Z_2 topological invariant defines the change in the TR polarization when the flux varies from 0 to $\varphi_0/2$ (see Fig. 1.7). If we take into account the spin-conserving case, the modification of the Kramers degeneracy is easy to highlight because the $\varphi_0/2$ flux insertion transfers "half" a spin up to the left and "half" a spin down to the right of the cylinder. Therefore, the eigenvalue of Sz, the Z-component of spin, associated with the end is logically divided by a factor 2, and so, as a consequence of the Kramers theorem, the degeneracy is automatically modified. Mitigating the Sz conservation (while preserving TR) forbids from labelling the states with Sz, even if the time variation in TR polarization subsists well defined. The presence of Kramers degeneracy is influenced by the number of local electrons that can be even or odd. Indeed, the thread of flux $\varphi_0/2$ through the cylinder acts as a "pump" for fermion parity. It is interesting to compare this interpretation with the edge state pictures in Fig. 1.8.

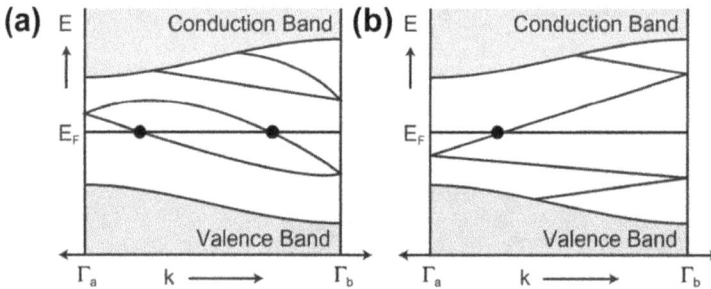

Fig. 1.8: Electronic dispersion between two boundary Kramers degenerate points. In (a) the number of surface states crossing the Fermi energy E_F is even, whereas in (b) it is odd (reprinted from [2], Copyright (2013), with permission from Elsevier).

If we consider the cylinder as a single large unit cell in the azimuthal direction, Fig. 1.8 details the discrete end state spectrum as a function of flux, with $k = a = 0$ corresponding to $\varphi = 0$ and $k = b = \pi/a$ corresponding to $= \varphi_0/2$. However, in case $\varphi = 0$, there are no partially occupied Kramers pairs, so that the many-body ground state is nondegenerate. Then at $\varphi = \varphi_0$ there will be a single half-filled Kramers pair, creating a Kramers degenerate many-body state, when the Z_2 invariant is non-trivial. Thus, flattening the cylinder into a Corbino disk, a flux $\varphi_0/2$ piercing a TI is associated with an odd fermion parity, but without a net charge. This is a typical example that highlights a situation where the spin and charge of the electron are not coupled.

1.5 The ideal case of graphene, which is not properly a topological insulator

Graphene is not what we could typically call a TI and we will explain the reason. In fact, we can state that graphene does not show a strong tendency to be a TI even if theoretically, at extremely low temperatures, it can show TI features. Anyway, the first study on the system constituted by the graphene helped scientists to elaborate a preliminary model of a potential TI. The main studies on this system have been developed by C.L. Kane at the University of Pennsylvania in Philadelphia (USA) [16]. But before discussing in detail the physics behind the QSHE, we need to briefly introduce the Haldane model which describes the QHE and allows understanding the origin of the TIs (QSHE) [2, 17]. Briefly this model says that if we consider the graphene and its band structure showing the typical Dirac structure, where the degeneracy at $q = 0$ (see Fig. 1.4) is protected by *inversion* and *TR symmetry*, respectively, P and T (one related to space and the other to time). In this model, the band is the typical 2D massless Dirac band structure. However, Haldane states that if we break on the two symmetries, we create a new term in the Hamiltonian, a mass which leads to the creation of a gap as shown in Fig. 1.9.

For example, the P symmetry is broken if two atoms in the unit cell are not equivalent, generating an insulating band structure (ordinary insulator or trivial insulator). The other way to break the TR symmetry is to apply a magnetic field. In this case, we generate a gapped band structure not typical of an ordinary insulator. In this case, we obtain a quantized Hall conductivity $\sigma = \pm e^2/h$, and we observe the QHE previously detailed. Up to now, we have considered a spinless system. Consider now the case of a system with a spin–orbit coupling (SOC). Firstly, we need to define what SOC specifically is. SOC is created by the interaction between the electrons and the charges in the nucleus of the atom, which is a relativistic interaction of a particle's spin with its motion inside a potential. Indeed, the spin–orbit interaction leads to a shift in an electron's atomic energy levels, related to electromagnetic interaction between the electron's magnetic dipole, its orbital motion, and the electrostatic field of the positive charge settled in the nucleus. The main effect is the splitting of the spectral lines. It is evident that more the atoms will be heavy (more charges in the nucleus) more important will be the interaction (considering that the electric field will rise) and so the effect: heavy atoms have strong SOC and that SOC is stronger in

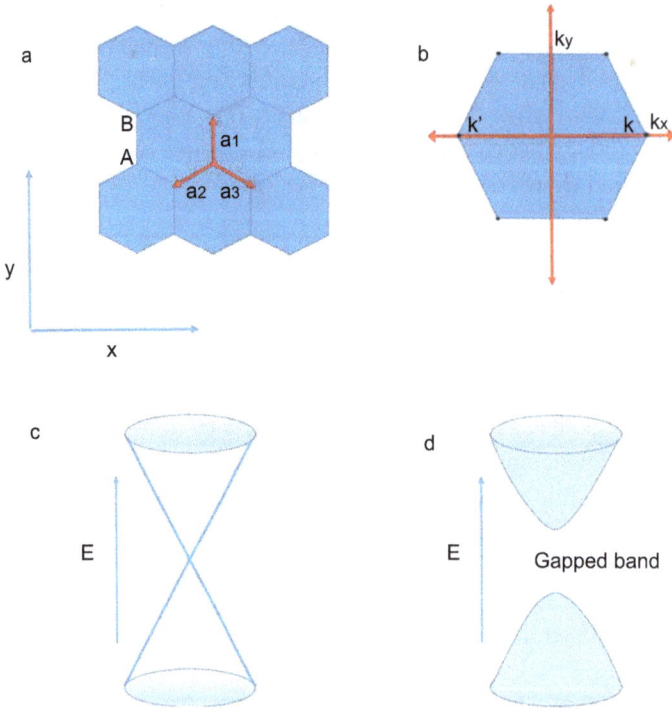

Fig. 1.9: (a) Graphene's molecular structure. (b) Graphene's Brillouin zone with two distinct corners K and K'. (c) Typical massless Dirac spectrum. (d) Typical massive Dirac spectrum after breaking symmetries with consequent energy gap.

case of electrons occupying p orbitals having nonzero angular momentum [18]. If we briefly summarize the three states (created by violating the symmetry and the third by adding the spin), we move from trivial insulator to QH state (with quantized conductivity) to TI (QSH states). This "transformation" is highlighted in Fig. 1.10.

Now, coming back to graphene as an ideal case, we will try to understand why taking into account the SOC we achieve a TI. Indeed when we add the SOC, we obtain a new mass term in the Hamiltonian (as in case of breaking the time or spatial symmetry). However, in this case this term depends on the value of spin which can do up or down (+1 or − 1). The main result that we do not have in this case is a single Hamiltonian but a double one depending on the value of the spin. If we take into account each one separately, we simply obtain two QH states that theoretically lead to QSH-quantized conductivity regime (as in the case of the QSH effect). The most important fact is that, considering that we have two copies of two QH states, we have two gapless edge states. The main difference is that even if the spin is not conserved, these two edge states are robust and they continue to exist (not like in case of QH states when we remove the magnetic field that

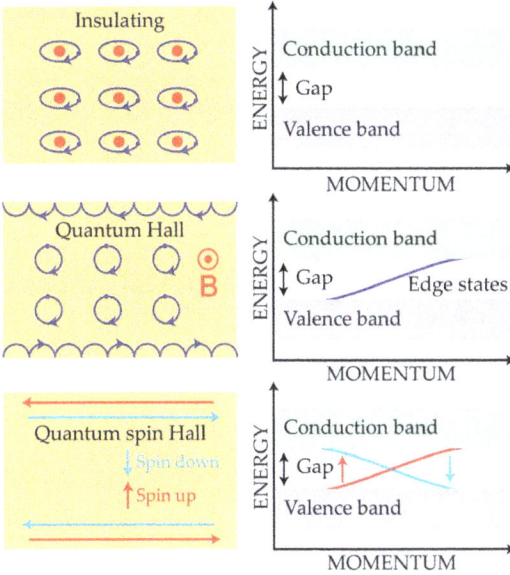

Fig. 1.10: This picture gives an overview of the three state that we are analysing: the trivial insulator state, the quantum Hall state with quantized resistance (the gap is crossed by conducting quantized state) and finally the quantum spin Hall state (topological insulator state) where we have spinful edge states with counter propagating spin-up, spin-down currents (reproduced from [18], with the permission of the American Institute of Physics).

generates the gap and the quantized conductivity). The edge states in case of the QSH effect have a "spin-filtered" property that can be translated with the fact that spin-up and spin-down propagate in opposite directions without losing coherence (thanks to the time invariance, as previously explained). Summarizing all the previous information lead us to two TR copies of the Haldane's model and so to a band gap that is pointed out in the following picture: it is likely to have two QH states superposed (see Fig. 1.11).

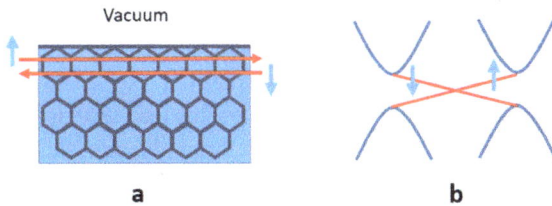

Fig. 1.11: Schematization of the QSHE and of the associated double band structure (a) the edge states in graphene have the spin-filtered property that up and down spins propagate in opposite directions. (b) The SOC leads to two Hamiltonians and so to two time-reversed Haldane's model generating two gapless "spin-filtered" edge states.

The main practical issue for graphene is that considering that the SOC is extremely low (understandably, because carbon is a light material), it is not possible to open a non-negligible gap in the material and to create the two gapless states at the edges. Moreover, considering that the gap will be nearly inexistent, electrons will be able to hop (thanks to thermal energy) in the different states, changing the configuration. Trying to take a more precise picture of the situation, it could be useful to analyse precisely what happens for a 2D material that seems to be particularly adapted to feature a measurable QSHE state. This is the case of stanene that can be called as the new "wonderful material" for 2D TI. We will talk about this material in the next section.

1.6 Stanene: the real wonderful material for beyond CMOS

Because of the success of 2D materials, a new class of material that previously existed only theoretically has been synthesized. We are talking about materials such as silicene [19] and germanene [20]. These last two are simply the 2D version of, respectively, silicon and germanium. These two materials did not show impressive characteristics from the physics point of view and are not potentially suitable TI. Actually, the main problem is related to the quite low SOC that does not allow opening a sufficient larger gap, as in case of graphene, stable for suitable environmental temperatures. As a direct consequence, it is not possible to observe the QSHE at room temperature for these materials. Moving to heavier atoms, in the same line of C, Si and Ge (all characterized by a honeycomb structure configuration of atoms), we finally meet tin. Stanene is the 2D form of tin. The name takes origin from the Latin name of tin, stannum. Considering that in hydrogen-like atoms, the SOC is proportional to Z [4, 21, 22], we can conclude that the SOC in stanene reaches a value around 5,000 larger than in graphene. For this reason, we can deduce that we can open quite easily a reasonable gap theoretically of around 0.1 eV in the bulk of this material. This apparent upper limit can be easily overcome through chemical functionalization of 2D materials, as suggested by Xu et al. in 2013 [23]. Thanks to chemical functionalization, scientists demonstrated that they could achieve a new class of materials and adapt the properties of 2D material films keeping topological features unaltered. This is also the case for 2D TIs and especially for 2D tin films. For example, we can firstly analyse the case of bare stanene. In this last case, it has been demonstrated that a low-buckled structure seems, as a results of modelling, to be the more stable for this specific material and not like in case of graphene-like structure [24]. This is due to the relatively weak π–π bonding between tin atoms in the lattice. The buckling enhances the overlap between orbitals and stabilizes the system. The same phenomenon happens for silicene which has a similar configuration. When we move to sp^3 hybridization, we observe the "decoration" of stanene. Therefore, in the decorated stanene, considering that the buckling of the tin nanosheet decreases, the Sn–Sn bond length slightly increases, and the equilibrium lattice constant grows. This is well highlighted in Fig. 1.12(c).

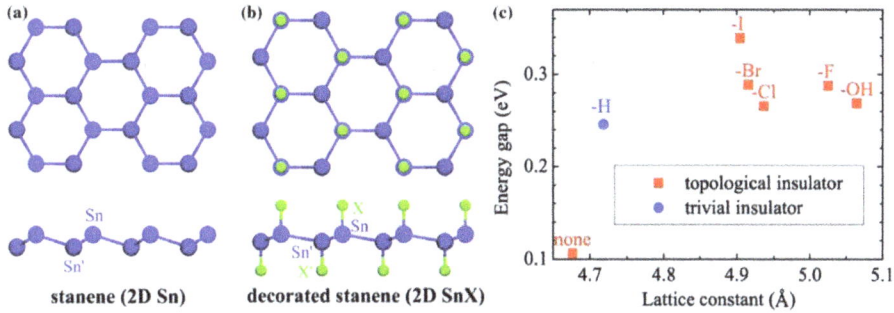

Fig. 1.12: (a) Stanene configuration with its typical low-buckled structure. (b) Decorated stanene in case of sp³ hybridization. (c) Energy gap as a function of the lattice for different decoration using different atoms (reprinted (abstract/excerpt/figure) with permission from [23], Copyright (2013) by the American Physical Society).

However, the most important effect is on the gap. Xu et al. took into consideration some common chemical functional groups such as –F, –Cl, –Br, –I, –OH and –H, to perform the theoretical calculation and verify if the TI characteristics were preserved. All but –H identified a TI state for the decorated materials. In fact, thanks to decoration, we can observe that the gap was enhanced through functionalization and reached a quite large value of 0.3 eV, as previously mentioned. Thanks to that we can imagine to achieve the QSH state at room temperature for these materials. This is without any discussion a major breakthrough considering the implementation of these materials for real applications in everyday life. In fact, for graphene, the strength of SOC is extremely weak and it gives rise to a very small gap of 0.02 meV [17]. This last can be considered negligible compared to the thermal energy $k_b T = 25$ meV at room temperature, leading to the electrical short-out of the edge states, becoming useless as conduction channels. Now we will try to explain briefly, what happens in case of functionalized stanene and specifically fluorinated stanene. In case of fluorinated graphene, the functionalization leads to a trivial insulator phase. Indeed, the low-energy physics at point K is completely changed by the filling of orbital π. This is not the case of fluorinated stanene where the functionalization drives to a gapless state at the point Γ if we do not take into consideration the SOC effect. In this last case, the SOC and the splitting of the energy state for the orbital $\boldsymbol{p}$, pushes the $\boldsymbol{s}$ state to a lower energy compared to the $\boldsymbol{p}$ state. In this last case, the degeneration of the $\boldsymbol{p}_{x,y}$ states is broken by the previously mentioned SOC. This inversion of the VB and conduction band because of the SOC interaction is exactly the same phenomenon that happens in

HgTe quantum wells, where the **p** and **s** states are inversed [25]. The main consequence of the inversion in point Γ is the build-up of a TI states as highlighted in Fig. 1.13.

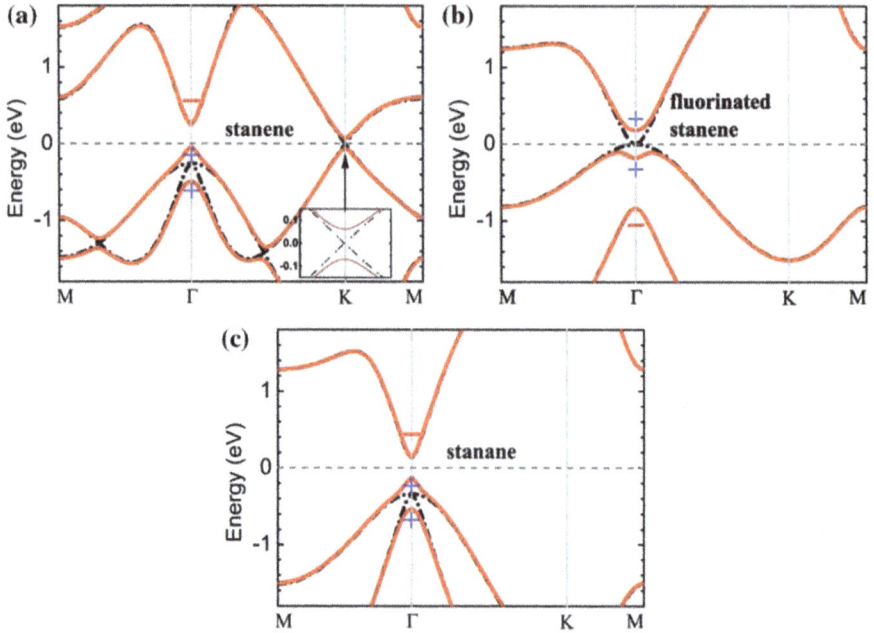

Fig. 1.13: Comparison between the band structure of stanene (a) and fluorinated stanene (b) before and after (c) taking into account the SOC effect (reprinted figure with permission from [23], Copyright (2013) by the American Physical Society).

Two-dimensional QSH insulators (a.k.a. topological insulators) with sufficiently large band gaps are mandatory for the development in real life of various innovative technologies especially for TE, as we will show in the next paragraphs. The main issue is how to move from dream to reality? More specifically, how can we grow these materials on suitable substrates and to continue observing the QHSE? This is exactly what happened for graphene that has shown extremely interesting properties from the very beginning as a free-standing layer. Scientists needed a lot of theoretical and experimental research to identify substrates that did not interfere with its properties, for example using hexagonal boron nitride. In case of stanene, through systematic density functional calculations and tight-binding simulations, it has been demonstrated by Wang et al. in 2016 [26] that if grown on an α-alumina surface it owned a substantial topologically non-trivial band gap (~0.25 eV) at the Γ point. The most interesting feature, also in the optics of implementation of these materials, is that even if stanene is atomically bonded to α-alumina, however, it is electronically decoupled from the substrate, bringing high structural stability and isolated QSH states to a large extent. This result is extremely important because it opens new routes on the exploration of TI keeping their features unaltered also at room temperature and so allowing their implementation for performing accurate deeper studies. Another interesting approach proposed by Xu et al. in 2018 [27] was to grow stanene on InSb substrates. In this case, researchers measured a gap of 0.44 eV, exploiting the Angle-resolved photo-emission spectroscopy (ARPES) technique. This value is larger than that already theoretically calculated for free-standing stanene. The main reason that led to this result was related to the electronic coupling between the Sn and InSb conduction band states. However, it has to be taken into account that only the single-layer case exhibits a gap. At two layers, the stanene gap was already filled in by InSb conduction band states. The main consequence was that these substrate states could act as an electrical short, making vanish the possibility to study the edge electronics in a suitable way. We can conclude that the stanene films on InSb are very promising candidates for QSH applications at room temperature and higher [27]. However, to accomplish this task, we need to develop extremely reproducible and reliable technological steps that will allow preserving the TI features also at ambient temperature. It is clear that the challenge is great considering that we are talking about artificial materials such as stanene and potentially plumbene, but the gain can be terrific in terms of applications.

1.7 Topological insulators and thermoelectric effect: the game changer?

It is commonly known that more than 60% of the energy used worldwide (in industrial plants, in transportation and so on) is completely lost, and the largest part through waste heat [28, 29]. For this reason, it has become a priority for governments and research institutes all around the world to work on the development of a new generation of materials that can in direct and reversible way convert this lost heat in a useable form of energy. In this context, an important position is kept by high-performance TE materials that can convert heat to electrical energy. A TE system is an environment-friendly energy conversion technology with the advantages of small size, high reliability, no pollutants and feasibility in a wide temperature range. Moreover, thermoelectricity does not imply the realization of moving parts, with consequent technological developments and maintenance issues, but mainly exploits the intrinsic properties of the materials. The main bottleneck to the exploitation on large scale of these materials is the efficiency of TE devices, which is not high enough to approach the Carnot efficiency describing the maximum thermal efficiency that a heat engine can achieve as permitted by the second law of thermodynamics. From this point of view, TI can be a real breakthrough and game-changer for the modern industry. A reasonable question can be raised: why are we talking about TI in this context? The answer is that these materials are extremely interesting for achieving a new generation of TE generators that will allow obtaining systems with a value of zT (also defined as the "TE efficiency") which could be one order or magnitude larger than the existing ones. In order to be more precise and pertinent, we can perform a review of the parameters that define the value of zT. The efficiency of TE materials is given by the dimensionless figure of merit $zT = (S2\sigma/\kappa)T$, where S, σ, κ and T are, respectively, the Seebeck coefficient, electrical conductivity, thermal conductivity and absolute temperature. To achieve excellent TE performances, a high zT value is expected, which means a large thermopower (the absolute value of the Seebeck coefficient that can be negative for negatively charged carriers, electrons, and positive for positively charged carriers, electron holes), a high electrical conductivity and a low thermal conductivity at the same time. In these terms, it seems easy to optimize the zT value. This is simply very far from reality, as we know. Indeed, the optimization of zT represents an extremely difficult task in material sciences and this is because these three parameters are strictly related and do not change independently their value. Now we have to understand why the same characteristics that are specific to TI are also important for TE materials. It has been demonstrated that many TIs are excellent TE materials. We can mention, for example, bismuth telluride (Bi_2Te_3), antimony telluride (Sb_2Te_3), bismuth selenide (Bi_2Se_3) and tin telluride (SnTe). Considering that one of the main features of TIs is the surface, for 3D materials, or edge conductivity for 2D materials, highlighting a metal-like behaviour, and an insulating one in the bulk (3D) or at the centre (2D), it seems that considering that the electronic transport is immune to backscattering by non-magnetic disorders and defects while it is not the case of phonon transport (bulk for 3D or centre of the materials for 2D), we can achieve the decoupling of electronics and thermal conductivity exploiting intrinsic properties without specific engineering of the material. Thus, the two kinds of transport can be effectively decoupled, realizing the "phonon glass, electron crystal" concept improving dramatically the TE performance in TIs [30]. We use the term of phonon glass considering that adding disorders and defects (non-magnetic ones) reduces the crystallinity of the materials making it more glassy (non-homogeneous). From the previous discussion, it appears that to achieve a strong enhancement of TE performance needs to reduce thermal conductivity while keeping electrical conductivity and Seebeck coefficient in high values. From this point of view, topologically protected boundary states in TIs working as conducting channels are very promising for this purpose, as they have lower physical dimension than the bulk states, showing excellent transport characteristics due to the absence of backscattering from defects with non-magnetic nature. Thanks to an ad hoc design of geometrical characteristics of the samples and adding disorders and defects (in the centre of the platform for 2D materials), the decoupling of the contributions of electrons (including both bulk and boundary states) and phonons can be enhanced further and TE properties of TIs improved.

1.8 Theoretical parameters influencing thermoelectric effect in topological insulators

As highlighted by Xu et al. in 2014 [31], we can also define the zT as a function of the cross-sectional area A and of the length of the material L. In this case, we obtain as follows:

$$zT = \frac{GS^2 T}{K} \qquad (1.8)$$

where S is the Seebeck coefficient, K is the thermal conductance obtained summing the contribution of electrons and lattice vibration ($K = kA/L$), G is the electric conductance ($G = \sigma A/L$) and T is the temperature. Indeed, we can see that the final value of the factor of merit will not depend on the geometrical size considering that A and L will be cancelled. However, we will show that the geometrical dependence emerges, thanks to the Seebeck coefficient value. How can we explain this dependence? To proceed we will focus on 2D TIs. In this case, a double band structure can be highlighted as a function of the geometrical configuration: gapless edge states with helical spin-momentum locking and gapped bands at the centre of the material as shown in Fig. 1.14. The spin helicity means that the spin-up electrons move only in one direction while spin-down electrons propagate in the opposite direction: the spin is locked to the direction of propagation of charges in the edges (spin-momentum helical locking). This phenomenon has been experimentally observed in recent ARPES experiments [32–35]. Actually, we can assume a linear dispersion for 1D edge states and a parabolic dispersion for 2D bulk states. They can be written as

$$E_1(k) = \pm \hbar k v - \Delta \qquad (1.9)$$

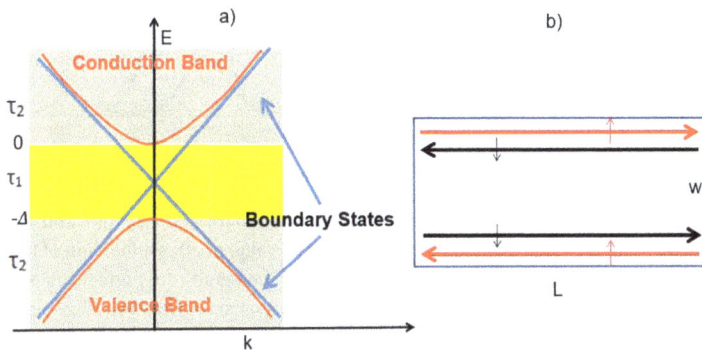

Fig. 1.14: (a) The band structure of a topological insulator with the boundary states. τ_1 and τ_2 are, respectively, the scattering time inside and outside the bulk gap. (b) Scheme showing a spinful configuration in case of 2D topological insulators.

$$E_2(k) = \hbar^2 k^2 / (2m^*) \tag{1.10}$$

In eq. (1.9), the plus and minus are related to the presence of two spins, **v** is the speed of the edge states and **k** the wave vector and **m*** is the effective mass. The minimum of the conduction band is selected as the reference level of energy. In Fig. 1.14, we represent the band structure and a detail of what happens at the gapless band edges. The values τ_1 and τ_2 are, respectively, the scattering times for edge states that are inside or outside the bulk gap. It is clear that intuitively inside the bulk gap considering the lower density of states (DOS) the value of τ_1 will be larger than τ_2.

The ratio between these two values will strictly be dependent on the geometrical characteristic of the configuration of the system. It is clear that the value can be enhanced in case of adding nonmagnetic defects or disorder in the system. But now, the most important point is to study the theoretical effect of these parameters on the value of the Seebeck coefficient and therefore on zT. We can see, for example, that the Seebeck effect associated with the boundary states is nearly zero in case of a length approaching zero and so when we are in the so-called ballistic limit. The main cause is that for these length values, the contributions of electrons and holes are opposite. Consequently, they make vanish their contribution considered that they are both thermally excited. However, for very large **L** value, we can see that the value of the Seebeck coefficient is strongly enhanced (see Fig. 1.15). In this case, we talk about diffusive limit.

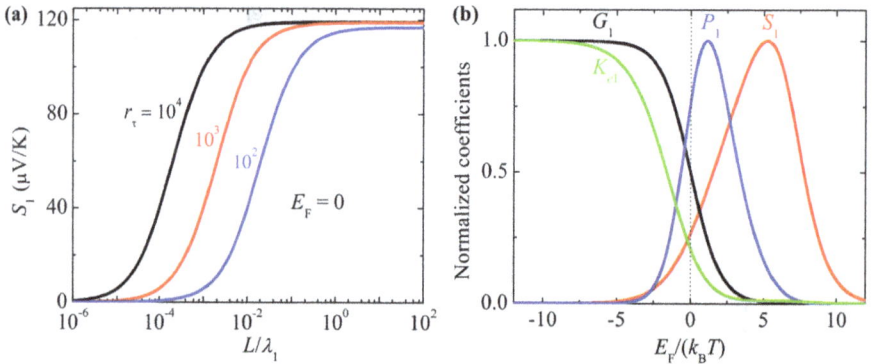

Fig. 1.15: (a) Seebeck coefficient as a function of the ratio between the length of the system and the free path of edge states (λ_1). (b) Main coefficients influencing the final value of zT as a function of the predefined Fermi level in case of diffusive transport and with $r_t = 10^3$. The results are normalized as a function of the maximum that can be reached (reprinted figure with permission from [41], Copyright (2014) by the American Physical Society).

Another important factor, as outlined by Xu et al. [32], is the definition of the Fermi level (E_F). We can observe that if the Fermi level is around $\sim 5k_bT$, the value of the Seebeck coefficient for a scattering time ratio, r_t, of 10^3 (which is the common value in case of HgTe quantum wells [44]) allows reaching a Seebeck coefficient value of 450 µV/K as shown in Fig. 1.10 [28]. Now, we would like to explain one very peculiar phenomenon occurring in TIs, which is the anomalous sign of the Seebeck coefficient. Briefly as known, the Seebeck coefficient corresponds to the response of electrons undergoing an external thermal gradient. As a function of the position of electrons above or below the Fermi level, we can observe a different behaviour heading to an opposite sign of S. When the TE transport is in large part led by the charges being above the Fermi level, S is negative. Otherwise, S is positive. In the case of a 2D TI, if the Fermi energy position is near the bulk conduction band and above the Dirac point, the charge carriers are expected to be electron-like from Hall measurements. However, for the Seebeck effect, the edge states above E_F are strongly scattered by defects/disorders due to the edge–bulk interactions (the DOS is larger than in the gap). In opposite, the edge states below E_F and within the bulk band gap are topologically robust and preserved from backscattering (and, logically, by the lower DOS). The contribution of the boundary states constitutes the most important contribution to the Seebeck effect, giving a positive (i.e. hole-like) S. The S sign will change, when the Fermi level is placed around the bulk VB. Therefore, taking into account the edge–bulk interactions, TI boundary states can simultaneously have sizable S and superior mobility: both are extremely important features for TE applications. It is well known that the position of E_F or of the chemical potential (μ) of electrons is a critical factor determining transport behaviours of a given TE material [36–38]. After the last paragraphs, it clearly appears that the Fermi level in TIs not only influences dramatically the transport behaviour of each channel (bulk or boundary channel) but doing that also determines their relative contribution to TE transport. However, we have to consider that TE transports of bulk and boundary states are optimized for the same range of values of E_F. We observe that zT of the boundary states is enhanced when E_F is outside the bulk band gap. In opposite, zT of the bulk states increases when E_F is within the bulk band gap. Moreover, if the Fermi level is inside the bulk band gap, carriers on boundary states dominate the TE transport in TI nanostructures. In opposite, the bulk states become dominating when E_F is outside the bulk band gap (e.g. high doping case). Therefore, to optimize the TE performances of the materials we need to identify a compromise. Calculations point out that the optimal E_F for large Seebeck coefficients is when its position approaches the bulk CB or VB, and so it is for the overall TE performance [39–42]. Different strategies can be adopted to change the value of the Fermi level, such as chemical doping and gating voltage, to define the best compromise to enhance the zT. However, a specificity of 2D TI has to be taken into account considering that if the boundary states are topologically protected to coexist with bulk states, zT, in this case, we can conclude that we are not in a situation dominated by the nanostructuration of the TI materials but mainly by their geometrical configuration. This is a peculiar characteristic of 2D TIs. In fact, the geometrical configuration greatly influences the hybridization of boundary states and the contribution to TE transport of

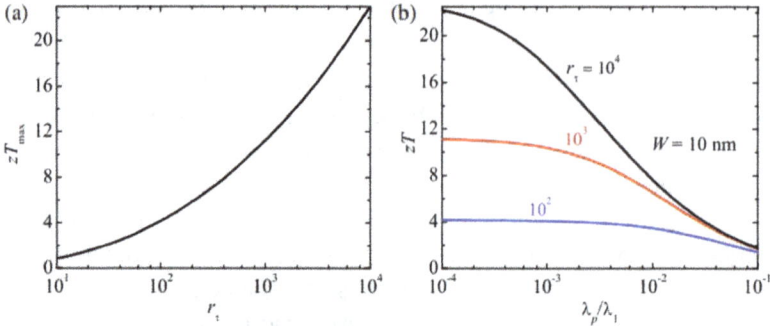

Fig. 1.16: (a) Maximum theoretical value for zT in case of 2D TIs as a function of the ratio between the scattering times of the edge states inside and outside the gap. (b) zT as a function of the ratios between the mean free path of phonons and of the inelastic mean free path of edge states (reprinted figure with permission from [41], Copyright (2014) by the American Physical Society).

boundary or bulk states. In a previous work of the same team, it was outlined that zT of 2D TIs could be tuned through the change of the transport length and ribbon width [32]: increasing the transport length and decreasing the ribbon width lead to the enhancement of the contribution of edge states, and so a larger zT. The contribution of edge states is strongly increased by the choice of a transport length larger than the inelastic scattering length (λ_i) and of a ribbon width twofold compared to the edge states one. This leads to the optimized geometry for total TE performance, achieving a huge enhancement of zT compared to the present state of the art ($zT < 3$, see Fig. 1.18).

From the previous simulation (Fig. 1.16), it is quite logic to see that when the r_τ ratio is high, the value of zT can increase in a dramatic way reaching, theoretically, 20 for 10^4. It is also logic that when the thermal conduction is lower and so the mean free phonon path smaller, the zT is larger and can reach nearly 12 (which is huge value considering the state of the art for this kind of materials) for r_τ reaching 10^3 (which is the common value in case of HgTe quantum wells [43]). In these simulations, Xu et al. considered a width (W) of 10 nm. A very interesting case to describe is the fluorinated stanene that we have described previously. Xu et al. analysed the dependence of zT in this material, one of the most promising 2D TIs. Indeed this material has already presented a non-trivial bulk gap of 0.3 eV which allows to highlight the typical features of TI at ambient temperature. Xu et al. supposed to have an inelastic mean free path for the edges and the width of the monolayer around 10 nm. The simulation is pointed out in Fig. 1.17.

We can see that for fluorinated stanene, we can theoretically reach an impressing value of 7, larger than the present state of the art values (see Fig. 1.18), with a ribbon width of 10 nm, as highlighted in Fig. 1.17. The Seebeck parameter reaches +300 μV/K for the same values.

We can observe that the largest value obtained using the bulk material does not reach 3. This is another demonstration that TIs can lead to a real shift in the

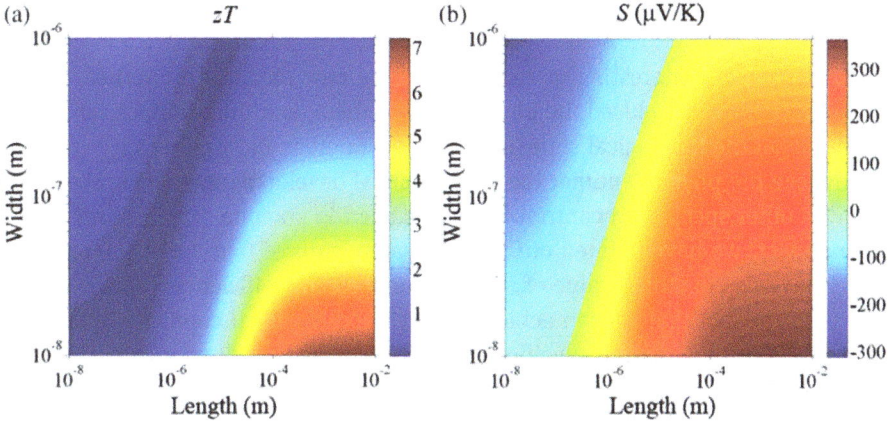

Fig. 1.17: (a) zT value as a function of the geometrical configuration of the nanoribbon. (b) Seebeck parameter as a function of L and W (reprinted figure with permission from [41], Copyright (2014) by the American Physical Society).

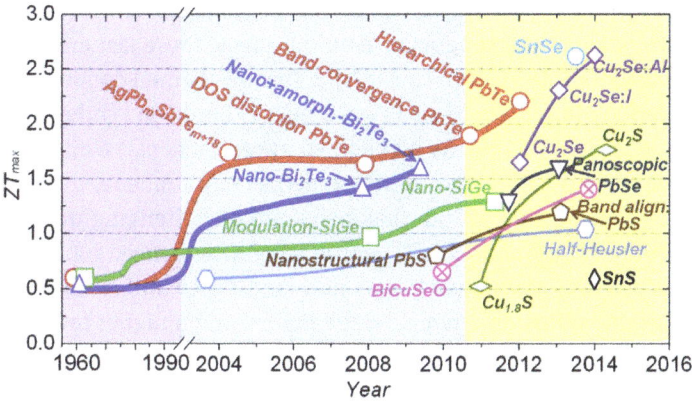

Fig. 1.18: zT of the current bulk thermoelectric materials as a function of year (reprinted figure with permission from [44], Copyright (2015) by the Elsevier).

paradigm if we will be able to manage to keep the peculiar properties of the materials developing a compatible process, for example, using a compatible substrate for growth that can preserve the TI features. This demonstrates the potential of 2D materials that exploiting their intrinsic exotic properties can show TE performances that can open large fields for innovation changing our everyday life.

1.9 Conclusions

TIs are extremely interesting materials featuring new physical properties. In this chapter, we have introduced the physics behind them, and the two main invariants used to label the topological state are the Chern number and Z_2. Our approach was to privilege the physics and not the mathematical developments that can be easily found in other specialized contributions. Considering potential applications with a huge impact, we have pointed out that the same features that are at the origin of the TIs give rise to a high value of zT (the thermoelectric efficiency). This is mainly related to the fact that we can achieve a very good electrical conductivity and exploiting doping or creating specific defects, we can reduce dramatically the main free path of phonons and so the thermal conductivity, achieving the so-called *phonon glass vision*. In 2D materials, the zT value can reach theoretically 20 and 7 in case of fluorinated stanene. This last is one of the most promising 2D materials considering its very high SOC that allows these materials to show TI features at ambient temperature (which is not doable, e.g. for graphene). Actually, thanks to SOC, theoretically it is possible to open a stable bulk gap for this material that can reach 0.3 eV preserving the materials from thermal excitations at ambient temperature. Thanks to the very high value of the conductance on the edges, these last are able to give a very huge contribution to zT, also considering the low thermal conduction of the bulk material. Very recently, plumbene (a.k.a Leadene) has been synthesized by Yuhara, Guy Le Lay et al. in 2019 [44]. As highlighted by researchers, "While theoretical studies predicted the stability and exotic properties of plumbene, the last group-14 cousin of graphene, its realization has remained a challenging quest." Considering that the SOC is proportional to Z^4, we can imagine enhanced topological features at ambient temperatures opening wide new landscapes in this research topic. The theoretical works in this field remain quite limited and a higher involvement of the scientific community could help identify the best directions to follow in terms of technology steps such as the identification of the best substrate to preserve the topology features or the best functionalization to stabilize the material enhancing the gap. A great part of the community tends to think that TIs have already been studied for thermoelectricity. This is a common mistake considering that the main works were limited to 3D materials, and the topological features were not adequately taken into account. Moreover, the physics of 2D TI materials is completely different. The 2D TI materials point out that with specific engineering on the materials in terms of geometrical dimensions we can tailor ad hoc platforms to enhance dramatically the zT and the Seebeck coefficient. This is completely new, and there are no teams in the world working specifically on this topic from growth to theory and tests. Indeed, we have observed in the last years that the scientific community tends to focus its work on theoretical aspects of topology without thinking sufficiently to real application of 2D materials, which own topological characteristics. We think that the 2D topological materials will be a real revolution. However, the attitude of the scientific community has to change and has to be focused in a more effective way the field of applications. Theoretical teams have to work in contact with experimentalists and with scientists able to grow these materials (with important technical challenges to overcome). Only thanks to that, we will be able to concretize our vision and to move from a simple fun physics object for producing papers in *Nature* and *Science*, to materials that will change our everyday life and will open new territories to explore in the field of physics.

References

[1]   https://www.britannica.com/biography/Felix-Klein
[2]   Topological Insulators, Volume 6, 1st Edition, Review Series Volume Editors: Marcel Franz Laurens Molenkamp Hardcover, Chapter One, Topological band theory and the Z2 invariant, C.L. Kane ISBN: 9780444633149 eBook ISBN: 9780444633187 Imprint: Elsevier Published Date: 15th November 2013, 352
[3]   https://mathworld.wolfram.com/Gauss-BonnetFormula.html
[4]   Do Carmo, M. P. Differential geometry of curves and surfaces. Prentice Hall, ISBN: 978-0-13-212589-7, 1976.
[5]   Adiabatic continuity, wavefunction overlap and topological phase transitions Jiahua Gu, Kai Sun. Phys. Rev. B. 94, 2016, 125111, https://doi.org/10.1103/PhysRevB.94.125111.
[6]   Xiao-Liang, Q., Zhang, S.-C. The quantum spin Hall effect and topological insulators. Phys. Today. 63, 2010, 1, 33. https://doi.org/10.1063/1.3293411.
[7]   Klitzing, K. V., Dorda, G., Pepper, M. New method for high-accuracy determination of the fine-structure constant based on quantized Hall resistance. Phys. Rev. Lett. 45, 1980, 494. https://doi.org/10.1103/PhysRevLett.45.494.
[8]   Rohrlich, D. Berry's Phase. In: Greenberger, D., Hentschel, K., Weinert, F., eds, Compendium of quantum physics, Springer, Berlin, Heidelberg, https://doi.org/10.1007 /978-3-540-70626-7_12, 2009.
[9]   in Zeitschrift für Physik 51, 165 1928.
[10] Berry, M. V. Proc. R Soc. Lond. A. 392, 1984, 45.
[11] Nakahara, M. Geometry, topology and physics. Bristol, Adam Hilger, 1990.
[12] Thouless, D. J., Kohmoto, M., Nightingale, M. P., den Nijs, M. Quantized Hall conductance in a two-dimensional periodic potential. Phys. Rev. Lett. 49, 1982, 405.
[13] Kubo, R. Statistical-mechanical theory of irreversible processes. I. General theory and simple applications to magnetic and conduction problems. J. Phys. Soc. Jpn. 12, 570–586. 10.1143/ JPSJ.12.570.
[14] Ryogo, K., Mario, Y., Nakajima, S. Statistical-mechanical theory of irreversible processes. II. Response to thermal disturbance. J. Phys. Soc. Jpn. 12, 1203–1211. 10.1143/JPSJ.12.1203.
[15] Laughlin, R. B. Phys. Rev. B. 23, 1981, 5632.
[16] Kane, C. L., Mele, E. J. Quantum spin Hall effect in graphene. Phys. Rev. Lett. 95, 2005, 226801. https://doi.org/10.1103/PhysRevLett.95.226801.
[17] Haldane, F. D. M. Model for a quantum Hall effect without Landau levels: condensed-matter realization of the parity anomaly. Phys. Rev. Lett. 61, 2015, 1988, https://doi.org/10.1103/ PhysRevLett.61.2015.
[18] Charles Day. Quantum spin Hall effect shows up in a quantum well insulator, just as predicted. Phys. Today. 61, 1, 2008);, 19, 10.1063/1.2835139.

[19] Vogt, P., DePadova, P., Quaresima, C., Avila, J., Frantzeskakis, E., Asensio, M. C., Resta, A., Ealet, B., Le Lay, G. Silicene: compelling experimental evidence for graphenelike two-dimensional silicon. Phys. Rev. Lett. 108, 2012, 155501. https://doi.org/10.1103/PhysRevLett.108.155501.

[20] Dávila, M. E., Xian, L., Cahangirov, S., Lelay, G. Germanene: A novel two-dimensional germanium allotrope akin to graphene and silicene. New J. Phys. 16, 9, 2014, https://doi.org/10.1088/1367-2630/16/9/095002.

[21] Landau, L. D., Lifshitz, E. M. Quantum mechanics: non-relativistic theory. Vol. 3. 3rd. Pergamon Press ISBN 978-0-08-020940-1, 1977.

[22] Pauling, L., Wilson, E. B. Introduction to quantum mechanics with applications to chemistry. New York, McGraw-Hill, 1935.

[23] Xu, Y., Yan, B., Zhang, H.-J., Wang, J., Xu, G., Tang, P., Duan, W., Zhang, S.-C. Large-gap quantum spin Hall insulators in tin films. Phys. Rev. Lett. 111, 136804 2013, https://doi.org/10.1103/PhysRevLett.111.136804

[24] Cahangirov, S., Topsakal, M., Aktürk, E., Sahin, H., Ciraci, S. Two- and one-dimensional honeycomb structures of silicon and germanium, Phys. Rev. Lett. 102, 236804 2009, https://doi.org/10.1103/PhysRevLett.111.136804

[25] Bernevig, B. A., Hughes, T. L., Zhang, S. C. Quantum spin Hall effect and topological phase transition in HgTe quantum wells. Science. 314, 5806, 2016, 1757–1761. 10.1126/science.1133734.

[26] Hui Wang, S. T., Kim, P. J., Wang, Z., Fu, H. H., Wu., R. Q. Possibility of realizing quantum spin Hall effect at room temperature in stanene/Al2O3(0001). Phys. Rev. B. 94, 2016, 035112. https://doi.org/10.1103/PhysRevB.94.035112.

[27] Cai-Zhi, X., Chan, Y.-H., Chen, P., Wang, X., Flötotto, D., Hlevyack, J. A., Bian, G., Sung-Kwan, M., Chou, M.-Y., Chiang, T.-C. Gapped electronic structure of epitaxial stanene on InSb(111). Phys. Rev. B. 97, 2018, 035122. https://doi.org/10.1103/PhysRevB.97.035122.

[28] https://ec.europa.eu/energy/sites/default/files/fi_ca_2020_en_a01_overview_eed_article_14.pdf

[29] Jouhara, H., Khordehgah, N., Almahmoud, S., Delpech, B., Chauhan, A., Tassou, S. A. Waste heat recovery technologies and applications. Therm. Sci. Eng. Prog. 6, 2018, 268–289. https://doi.org/10.1016/j.tsep.2018.04.017.

[30] Nolas, G. S., Sharp, J., Goldsmid, H. J. The phonon – glass electron-crystal approach to thermoelectric materials research. In: Thermoelectrics. Springer series in materials science. Vol. 45, Springer, Berlin, Heidelberg, 2001.

[31] Xu, Y., Gan, Z., Zhang, S. C., Enhanced thermoelectric effect performances and anomalous Seebeck effect in topological insulators. Phys. Rev. Lett. 112, 226801 2014, https://doi.org/10.1103/PhysRevLett.112.226801

[32] Hsieh, D., Xia, Y., Qian, D. et al. A tunable topological insulator in the spin helical Dirac transport regime. Nature. 460, 2009, 1101–1105. 10.1038/nature08234.

[33] Xia, Y., Qian, D., Hsieh, D. et al., Observation of a large-gap topological-insulator class with a single Dirac cone on the surface. Nat. Phys. 5, 2009, 398–402. https://doi.org/10.1038/nphys1274.

[34] Chen, Y. L., Analytis, J. G., Chu, J. H., Liu, Z. K., Mo, S. K., Qi, X. L., Zhang, H. J., Lu, D. H., Dai, X., Fang, Z., Zhang, S. C., Fisher, I. R., Hussain, Z., Shen, Z. X. Experimental realization of a three-dimensional topological insulator, Bi2Te3. Science. 2009, 10;325, 5937, 178–181, 10.1126/science.1173034.

[35] Hirahara, T., Sakamoto, Y., Takeichi, Y., Miyazaki, H., Kimura, S.-I., Matsuda, I., Kakizaki, A., Hasegawa, S. Anomalous transport in an n-type topological insulator ultrathin Bi2Se3 film. Phys. Rev. B. 82, 2010, 155309. https://doi.org/10.1103/PhysRevB.82.155309.

[36] Snyder, G., Toberer, E. Complex thermoelectric materials. Nat. Mater. 7, 2008, 105–114. https://doi.org/10.1038/nmat2090.

[37] Pei, Y., Wang, H., Snyder, G. J. Band engineering of thermoelectric materials. Adv. Mater. 24, 2012, 6125–6135. https://doi.org/10.1002/adma.201202919.

[38] Ioffe, A. F. Semiconductor thermoelements and thermoelectric cooling. Infosearch Press, 1956.

[39] Murakami, S., Takahashi, R., Tretiakov, O. A., Abanov, A., Sinova, J. Thermoelectric transport of perfectly conducting channels in two- and three-dimensional topological insulators. J. Phys.: Conf. Ser. 334, 2011, 012013. 10.1088/1742-6596/334/1/012013.

[40] Takahashi, R., Murakami, S. Thermoelectric transport in topological insulators. Semicond. Sci. Technol. 27, 2012, 124005. https://doi.org/10.1088/0268-1242/27/12/124005.

[41] Xu, Y., Gan, Z., Zhang, S. C. Enhanced thermoelectric performance and anomalous Seebeck effects in topological insulators. Phys. Rev. Lett. 112, 2014, 226801. https://doi.org/10.1103/PhysRevLett.112.226801.

[42] Takahashi, R., Murakami, S. Thermoelectric transport in perfectly conducting channels in quantum spin Hall systems. Phys. Rev. B. 81, 2010, 161302. https://doi.org/10.1103/PhysRevB.81.161302.

[43] Topological Insulators, Volume 6 1st Edition 5.0 Review Series Volume Editors: Marcel Franz Laurens Molenkamp Hardcover, Chapter three, Model and materials for topological insulators, C. Liu, S. Zhang, ISBN: 9780444633149 eBook ISBN: 9780444633187 Imprint: Elsevier Published Date: 15th November 2013, 352

[44] Yuhara, J., He, B., Matsunami, N., Nakatake, M., Lay, G. L. Graphene's latest cousin: plumbene epitaxial growth on a "nano waterCube". Adv. Mater. 1901017, 2019, https://doi.org/10.1002/adma.201901017.

2 Magic angle: twisting graphene layers to create superconductors

2.1 Introduction

The history of the research on the magic angle for graphene and more generally for 2D materials is only at its early stage; however, incredible new horizons for research are emerging, thanks to its discovery. The word "twistronics" has been invented by Carr et al. in 2017 [1] to define the new science dealing with this kind of phenomena for 2D nanomaterials. Magic angle and twistronics are hot topics for research. Each day we can observe the publication of a large quantity of new scientific papers shedding light on the theoretical and experimental results concerning this field. In this contribution, we will try to explain, in a comprehensible way, the basics and the most important aspects related to research on magic angle. We will try to understand if this topic can potentially lead to a new generation of devices or if it is simply a bubble created by theoreticians for publishing high-rated papers. The reader will be able to make her/his own opinion. We will try to introduce each concept in a simple and intuitive way also considering the complexity, and the quite limited knowledge, of the theoretical science behind. But, as a first point, what is in a simple and clear way the magic angle? Very briefly, we can say that it is the twisting angle between two stacked layers of graphene, which emerges as superconductivity states. Magic angle existence was theoretically predicted and experimentally proven 6/7 years after. When two stacked layers of graphene, or 2D materials in general, are twisted, we create a Moiré superlattice having peculiar physical features. A Moiré superlattice is simply a system obtained by the superposition of two patterns, for example the hexagonal configuration of atoms in graphene. The two main configurations of superposition of graphene hexagonal pattern, before twisting them, are AA and AB (see Fig. 2.1). In the first one, the two layers are exactly coincident, and in the second one, they are translated one from the other by the distance of two carbon atoms of the lattice. Starting from these two configurations, twisting the two patterns/layers, we can modulate the electronic band structure leading to transport properties such as unconventional superconductivity [2] or insulating behaviour of the stacked structures.

In the next sections, we will introduce the main concepts related to the magic angle physics. We will introduce superconductivity, and more specifically unconventional superconductivity, and will explain how scientists were able to move from the prediction of the magic angle in 2011, disclosed by the work of Allan McDonald and co-workers [3], to the experimental evidence achieved by the team of Pablo Jarillo-Herrero at MIT [2, 4] with two papers published in *Nature* the same day, the 5th of April 2018. Finally, we will show which can be the perspectives, or at less we will try, in terms of applications for this phenomenon in the next years, trying to achieve realistic conclusions.

https://doi.org/10.1515/9783110656336-003

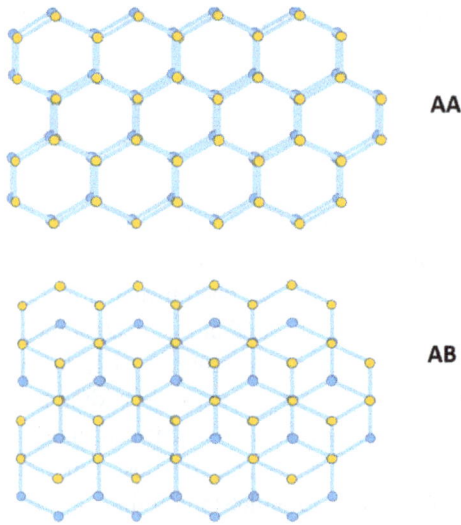

Fig. 2.1: Stacking orders for two graphene monolayers. In the AA case, the atom positions are superposed without any translation. In the AB case, we have a translation only in the *y*-direction of half the cell dimension.

2.2 A brief history of superconductivity

Superconductivity is one of the key concepts to explain the phenomena related to the magic angle. Superconductivity is the set of physical properties observed in specific materials, wherein electrical resistance goes abruptly to zero at the so-called critical temperature (T_C) and from which magnetic flux fields are expelled. One of the main features of materials defined as superconductors is that their behaviour is not like in case of common conductors such as metals where the conductivity changes linearly, increasing, as a function of the reduction of the temperature. To better understand the context, it is better to adopt an "historical approach" and to present the first scientific evidences. The sudden reduction of the resistance in superconductors was discovered in 1911 by the Nobel physics laureate, the Dutch physicist Heike Kamerlingh Onnes [5, 6], who was performing a study of the physics of materials at extremely low temperatures. On 8 April 1911, after testing for electrical resistance in a sample of mercury at 3 K, Onnes observed that the resistance was nearly zero: this was the first observation of a superconductor [7]. To be more precise, superconductivity is characterized by the Meissner effect, from the name of the German physicist Fritz Walther Meissner who discovered it in 1933 [8], which consists in expelling the magnetic field lines during the material transition into a superconducting state under the T_C. This is not coherent with the classical physics view that tended to make the superconductivity coincide to an ideal perfect conductor state. Actually, physicists needed half a century to grab the signification of conventional superconductivity, thanks to the theory developed in collaboration by John Bardeen, Leon Cooper and John R. Schrieffer, called BCS theory, in 1957 [9]. Thanks to that, they received the Nobel Prize in physics in 1972 [10]. Intuitively, the theory details superconductivity as a microscopic physical effect linked to the condensation of Cooper pairs. But what is a Copper's pair? The Cooper pairing is a quantum effect that, however, can be described using a simplified classical explanation [11] to understand in an easier way and rapidly its meaning. An electron in a metal can be associated with a free particle. The electron is logically repelled from other electrons because they have the same negative charge. In opposite, it is also subject to the attraction of the positive ions making up the lattice of the metal. This attraction gives rise to a deformation of the ion lattice because the positive ions move to some extent in the direction of the electron. Therefore, this increases the positive charge density of the lattice near the electron. This positive charge, consequently, can bring to itself other electrons (see fig. 2.2). At long distances, this attraction between electrons, due to the change of the position of positive ions, can be stronger than the electrons' repulsion related to their negative charge and couple them. If we want to give a quantum mechanical explanation, we can say that the effect is mainly due to electron–phonon interactions and that the phonon can be associated with the flow of the assembled positive ions in the lattice [12].

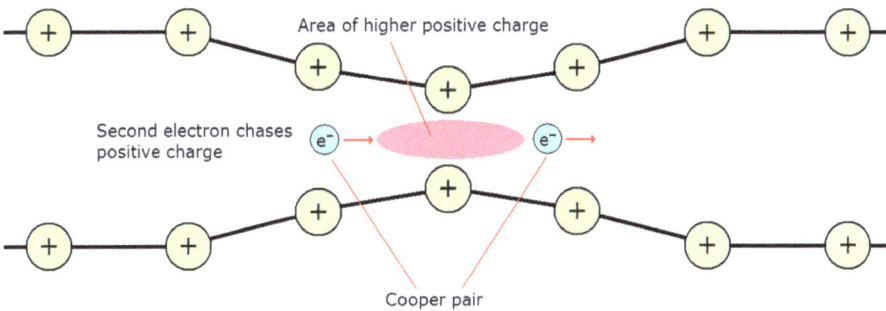

Fig. 2.2: Schematization of Cooper's pairing of electrons [13].

Considering that in a superconductor, we can observe a large quantity of such coupled electrons. These last overlap very strongly and finally they form a "whole set". Indeed, the breaking of one pair will automatically have an influence on the energy of the whole and not just on a single particle or coupled charge. The energy necessary to break any single couple is associated with the whole energy required to free up the pairs (or more than just two electrons). Because the charge coupling is at the origin of the energy barrier increase, the interactions with oscillating atoms in the conductor (quite reduced at these low temperatures) are not sufficient to put at risk the coherence of assembled charges, or any individual couple composing the condensate.

Thanks to that, the electrons stay paired together, and the electrons flow as a whole (the current through the superconductor) without experiencing resistance. This is the genesis of the superconductivity. The condensate collective behaviour is one of the main features necessary to reach a superconductivity state. A major shift in the paradigm for superconductivity was reached, thanks to the discovery by Georg Bednorz and Alex Müller working at the IBM Laboratory in Zurich, Switzerland, in 1986, of cuprate–perovskite ceramic materials that showed a critical temperature above 90 K (−183 °C) [7, 14]. This result was theoretically impossible to achieve following the conventional theory of superconductivity. For this reason, these materials were denominated "high-temperature superconductors" and the phenomenon named "unconventional superconductivity". Indeed, this was a real shift in the paradigm considering that thanks to commonly available coolant liquids, such as nitrogen that boils at 77 K, and the existence of superconductivity at higher temperatures could be proved by setting up many experiments and applications. These last would have been extremely difficult to build up at lower temperatures near the absolute zero. One year after, the discovery was confirmed by physicists [15] in the United States who found a material in the same class becoming superconducting at 93 K. To be sincere, the results of these experiments were, as usually happen for great discovery, the fruit of serendipity. Indeed, Bednorz and Müller's work was not focused on metals but on insulating materials called copper oxides than the defined cuprates [16]. In particular, they were trying to understand what happens when cuprates are doped and so when other elements such as lanthanum or barium were added into the parallel planes of copper and oxygen that comprise its structure. They observed that the newly added atoms contributed to free up the outer electron of some of the copper atoms, which then can move through the lattice. Consequently, if the cuprates were subject to a sufficient low temperature, depending on the level of their doping, the electrons could move without any constraint, and the material logically becomes a superconducting one. The most pertinent theory, which is the most supported by scientists, called spin fluctuation, suggested in 1991 by Philippe Monthoux of the University of Edinburgh (UK), Alexander Balatsky from Los Alamos National Laboratory in New Mexico (USA) and David Pines from the University of Illinois–Urbana Champaign (USA) [17], states that the material is not doped, and cuprates are stuck into an ordered state that can be defined as an antiferromagnetic. Therefore, the outer electron on each copper atom is aligned in such a way that its spin is opposite to that of its neighbour. Each line of electrons shows an alternating spin value for each charge composing it (up, down, up, down and so on). The main consequence is the production of a magnetic field that sticks the electron in place. This is not true for doped cuprates because the atoms added through doping do not allow respecting this ordered state and so the spin can switch. The main effect is that when we have a single moving charge this one can create a sort of deformation in the spin texture of the whole assembly of charges, a phenomenon that has clear analogy with the description of BCS theory of superconductivity. The deformation in the spin texture can bring electrons together and associate them in order to obtain Cooper pairs realizing a superconducting state.

2.3 The discovery of the magic angle: the prediction

After briefly introducing a brief history of superconductivity and some basic concepts, now we can start talking about the discovery of the magic angle and of its consequences on condensed matter but also on science in general. Some years ago, more precisely in 2011, at the University of Texas, Allan H. MacDonald, a theoretical physicist exploiting quantum mathematics and computer modelling to study 2D materials, working with one of his postdoctoral researchers, Rafi Bistritzer, discovered something of completely unexpected [3]. Indeed, they were working on stacked monolayers of graphene twisted one relative to the other by quite small angles (<5°). Their primary objective was to identify a simple example to obtain a computable problem reducing the complexity of the system compared to models implying a multitude of stacked layers (such as in case of graphite). The extremely surprising results that were highlighted by the simulation for different twisting angles (<2°) pointed out that the two layers became more strongly coupled and that the Dirac velocity crosses zero more times as the twist angle moved to lower values. They observed that this happened first at a twisting angle of 1.05° and this, as previously said, at the same time that the Dirac velocity went to zero. The main reason for this result was correlated to the flattening of Moiré band (band created as a consequence of pilling up two different patterns) and to the dramatic rising of the effective mass. Because of the flattening of the bands, they observed a peak in density of states (DOS). In Fig. 2.3, we can observe the difference between a twisting angle of 5° and two angles <2° as obtained by McDonald et al. in 2011.

In Fig. 2.3, we observe that in case of the two smaller angles, the DOS peaks because of the flattening of the bands. Therefore, as highlighted, at some specific angles, electrons behave in a strange and extraordinary way: they are suddenly more than 100 times slower. This result was only in part a confirmation of the pioneering work of 2007 by Lopes Dos Santos, Castro Neto and co-workers, which already had predicted the drastic reduction of the speed of charges for specific small angles [18].

As suggested by McDonald, the velocity tends to zero (see fig. 2.4) at the first magic angle because it is in the process of changing sign, and the eigenstates at the Dirac point are a coherent combination of components present in the two layers that have the velocity of opposite sign. However, McDonald, in the seminal paper of 2011, did not dare to state that his theoretical work could predict the existence of superconducting states at specific angles because of the strong interaction of slow electrons in the two twisted layers. This is the main limit of this work. For this reason, this extremely important study, which opened incredible new fields of research in condensed matter physics, was initially nearly ignored. Another reason of the underrating is mainly related to the technical hurdles to overcome in achieving a physical example of

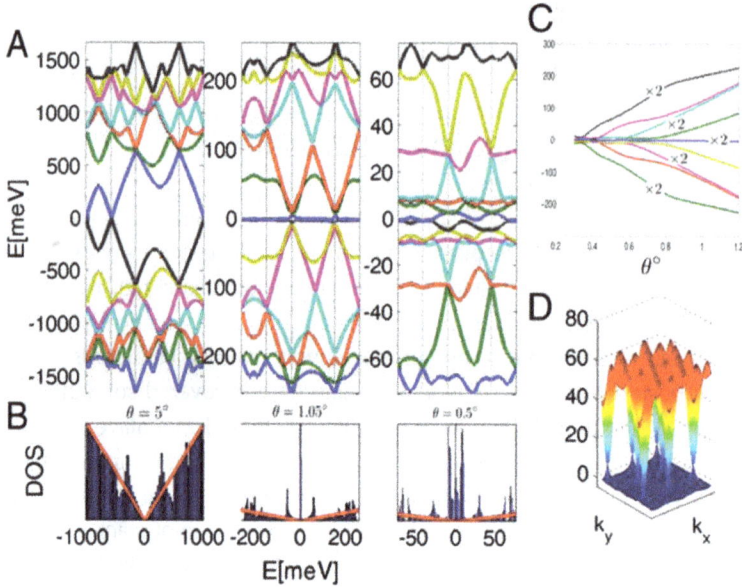

Fig. 2.3: Moiré bands for different twist angles. (A) For the bands closest to the Dirac point for $w = 110$ meV, and $\theta = 5°$ (left,), 1.05° (middle) and 0.5° (right). (B) DOS in the three cases [3]. (C) Energy as a fucntion of the twist angle. (D) Full disperion of the flat band at 1.05° (reproduced from [3], with the permission of *PNAS*).

this kind of system and to perform extensive experiments to verify the theoretical predictions. We can affirm that this paper maybe too advance for its times. After 5 years of intensive work, however, McDonald and co-workers were able to fabricate suitable structures in 2016 when finally a technique to stack two layers of graphene in a reliable way was optimized [19] (see Fig. 2.5).

In this pioneering work of 2016, the twisted graphene bilayer samples were fabricated by sequential graphene and hexagonal boron nitride (hBN) flakes picked up using a hemispherical handle substrate (see Fig. 2.5a), allowing to move an individual flake from a substrate while leaving the others in its immediate proximity intact. The graphene flake was split into two parts that then were collected using hBN flakes stuck to the hemispherical handle. Finally, between the first and the second graphene flake pick-up, the substrate was rotated by a specific small (0.6–1.2°) angle (see Fig. 2.5b and c) that can be controlled to 0.1° accuracy. The extremely smart approach of researchers was to use the same graphene flake, having the same crystal grain and so crystal axes were aligned, achieving a precise AA stacking of the two monolayers. Indeed, the substrate rotation allowed achieving a very precise twist between the two graphene layers (Fig. 2.5c) creating a Moiré crystal (Fig. 2.5d) in a reproducible way. To achieve a functional device that could be suitably tested, it was necessary to encapsulate the two layers in an hBN dielectric [20] and defining top-gate and edge metal contacts (see

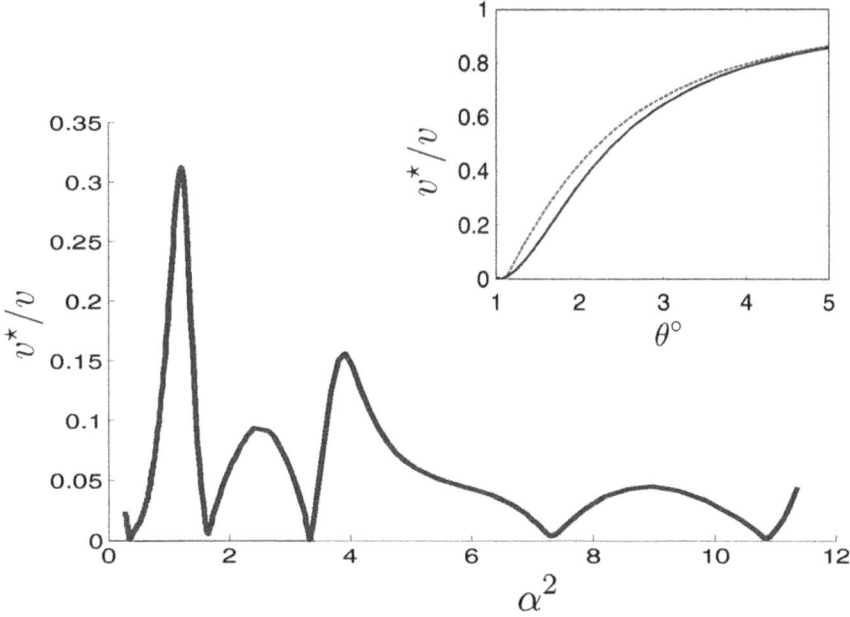

Fig. 2.4: Renormalized Dirac-point band velocity. The band velocity of the twisted bilayer at the Dirac point v^* is plotted versus α^2, where $\alpha \frac{1}{4} w/vk\theta$ for $0.18° < \theta < 1.2°$. The velocity goes to zero for $\theta \approx$ 1.05°, 0.5°, 0.35°, 0.24° and 0.2°. (Inset) The renormalized velocity at larger twist angles. The solid line and dashed line correspond, respectively, to numerical results and to analytic results (reproduced from [3] (2011), with the permission of *PNAS*).

Fig. 2.5e) [21]. The set-up shown in Fig. 2.5(d) allowed performing measures of the four-point conductance as a function of the top-gate bias to evaluate the effect of temperatures (T) (see Fig. 2.5f). The data pointed out a local conductance minimum when the carrier density (n) approached zero (charge neutrality) that is similar to the minimum seen in simple gated graphene samples. More surprisingly, they observed two pronounced conductance minima for a top-gate bias of ± 2.2 V, corresponding to a carrier density of $\pm 2.5 \times 10^{12}$ cm^{-2}. All conductance minima logically had the tendency to disappear increasing temperature, and vanished at temperatures above 80 K. This phenomenon observed is different from what was anticipated considering the density dependence of the conductance expected in either Bernal stacked [22–25] or large-angle twisted bilayer graphene (TBG) [26]. In fact, this effect is more similar to what happened to the conductance of graphene closely aligned with an hBN substrate [27–29]. Scientists explained the existence of conductance minima filling the first two bands of states produced by the Moiré crystal with electrons or holes. However, not intuitively, the conductance minima were stronger than what usually happened for graphene on hBN and took place at different values of carrier density per Moiré period, highlighting a strong temperature dependence. This phenomenon seemed to point out a change in the band diagram with the opening of a gap for charged excitations.

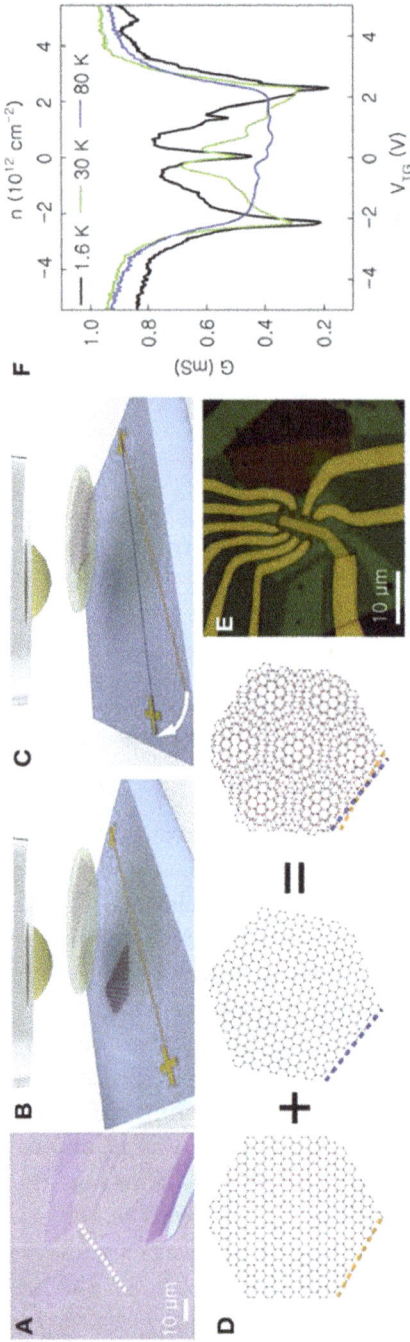

Fig. 2.5: Small twist angle (STA) bilayer graphene. (a) Optical micrograph showing a single graphene flake that after that splits into two sections as highlighted in the figure. (b, c) Handling of the two graphene flake sections. The second flake fragment is rotated to achieve a specific twist angle. (d) The Moiré pattern formation of the two superposed graphene flakes. (f) Conductivity of the bilayer graphene as a function of temperature (reproduced from [19] (2016), with the permission of *PNAS*).

2.4 The experimental results and evidence of unconventional superconductivity

Other scientists, not directly working with McDonald, showed a strong interest for the seminal work of 2011. Indeed, Pablo Jarillo-Herrero at MIT, who had already been working on this topic for 1 year, decided to focus his team's work on the realization of suitable samples able to confirm the prediction of McDonald and co-workers. These efforts did not produce significant results immediately. This seemed to confirm the lack of confidence in this topic mainly related to the extremely high challenge in finding suitable samples. Finally, in 2018, 2 years after the first samples tested by McDonald and co-workers, Jarillo-Herrero's team was able to set up a system of layered graphene twisted by 1.1° showing superconductivity at a surprisingly high temperature. This was and stays one of the major breakthroughs for science in the recent years. We can comfortably states that MIT scientists were able to take the lead in the field of twistronics, thanks to this experimental demonstration. Actually, Jarillo-Herrero and co-workers were able to go well beyond the work of McDonald et al. of 2011, intuitively predicting the existence of superconductive states. This was a giant leap of his work compared to the previous ones. As recognized by McDonald recently, "Certainly superconducting is the thing you most hope to see, but I didn't have the nerve to predict it" [30]. Two seminal papers were published immediately after the disclosure of the results, exactly 5th April 2018 [31, 32]. Jarillo-Herrero and co-workers were able to build up a van der Walls structure with two stacked layers of graphene with a structure very similar to the work of 2016. They succeeded in twisting the two layers at very small angles (around 1°). They observed, as previously announced by McDonald, the flattening of the band near the Fermi level because of the strong correlation between the two layers. Scientists of MIT observed that this band showed an insulating behaviour when half filled. This observation was not expected in case of any interactions between electrons. The main phenomenon at the origin of this observation is mainly due to the strong DOS inside the flat bands. One of the consequences is that the Coulomb interactions are largely more important than the kinetic energy of charges considering that the Fermi velocity tends to zero. This case can be assimilated to a "Mott-like insulator" [33], whereas Mott insulators mean a class of materials that should conduct electricity under conventional band theories, but that are, in fact, insulators because of the strong electron–electron interactions, not considered in the conventional band theory. In case of two stacked layers of graphene, the interlayer hybridization induces nearly flat low-energy bands. Upon electrostatic doping of the material away from correlating insulating states (so changing the filling of the flat bands), Jarrillo-Herrero and co-workers observed superconductivity states, to be more precise "unconventional superconductivity", which were similar to that already highlighted in cuprates [34]. With the term "cuprates", we generally define a class of materials that contain anionic copper complexes. Indeed, the name comes from cuprum which is the Latin word for copper. As previously highlighted, we would like to stress the fact that we are dealing with "unconventional conductivity" considering that this behaviour could not be simply explained using the common theory base on electron–phonon weak interaction. The most important results of the pioneering work of the team of MIT was the demonstration that only using two stacked layers of graphene, and so a simple networks of carbon atoms, they were able to fabricate a superconductor exploiting the generated Moiré pattern highlighted when a superlattice modulation is achieved applying a twist angle between the two monolayers. Compared to the previous study on superconductivity, it was no more necessary to grow different samples with different doping. In the case of the twisted graphene bilayer, only applying a bias, scientists were able to explore different charge configurations. All these results pushed André Bernevig, one of the pioneers of topological matter physics, to state that *these systems constituted an incredible playground for scientists to study the correlation states of matters and their relationship with superconductivity in a relatively easy way* [30]. Coming back to the Moiré superlattice, it takes origin from the hybridization between the two band structures of the monolayers due to the hopping that creates important modifications to the low-energy band structure depending on the stacking order (AA or AB, see Fig. 2.1). However, to be more precise, and to have a better overview of the situation, we can observe the following picture from one of two seminal papers of Jarillo-Herrero and co-workers [32].

In Fig. 2.6(d), we can see the how the mini-Brillouin zone linked to the Moiré superlattice is based on the difference between the two K (or K') which are the wave vectors for the two layers. The Dirac cones near one of the two valleys K or K' mix through the interlayer hybridization as highlighted in Fig. 2.6(e). The main consequence of this hybridization is the creation of gaps opened near the Dirac point and the renormalization of the Fermi velocity in these points. As previously pointed out, the calculated magic angles are specific twist angles between two graphene monolayers where the Fermi velocity goes to zero and the related energy bands, as outlined in Fig. 2.6(c), are flattened. This is very difficult to explain from an analytical point of view. Intuitively, it happens when the hybridization energy (see Fig. 2.6f) becomes comparable or larger than the value of the difference of the energy of the hybridized cones and the intersection of the two Dirac cones related to the initial monolayer. Looking more carefully to Fig. 2.6(f) we can see that the value of the energy of the hybridized cones compared to the intersection point can be defined as 2 W. W is defined as the interlayer hopping energy. The energy difference between the bottom of the hybridized cone and the Dirac point of the two monolayers is $\hbar v_0 k_\theta$, where v_0 is the Fermi velocity and k_θ is defined as the momentum displacement of the Dirac cones. This last is done by

$$k_\theta = G_K \theta \tag{2.1}$$

where $G_K = 4\pi/a$ and is defined as the magnitude of the wave vector Γ–K for graphene. a is the lattice constant of graphene which is 0.246 nm. Now if we took the

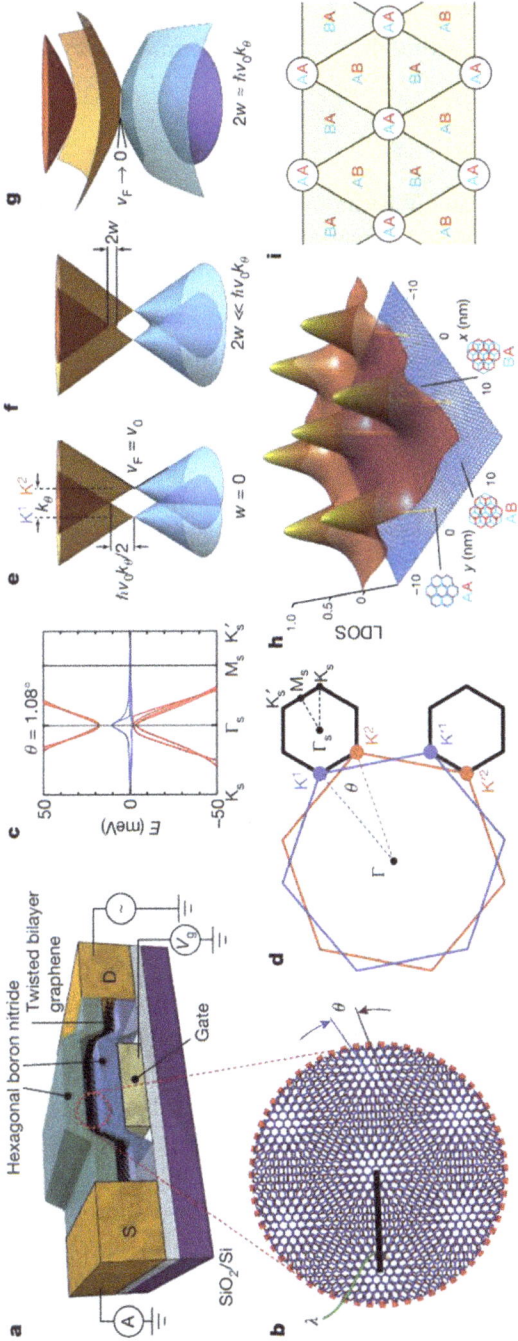

Fig. 2.6: Electronic band structure of the twisted bilayer graphene (TBG). (a) Configuration of the device, (b) the Moiré pattern of the stacking, (c) band energy E corresponding to $\theta = 1.08°$ with flat band in blue, (d) the mini-Brillouin zone of the superlattice, (e–g) illustration of the effect of interlayer hybridization and (h) normalized local density of states (LDOS) calculated for the flat bands with $E > 0$ at $\theta = 1.08°$. The electron density is strongly concentrated at the regions with AA stacking order and mostly depleted at AB and BA stacked regions. (i) Scheme of the stacking order (reprinted with permission from Springer Nature Customer Service Centre GmbH, Springer Nature [32], Copyright Springer (2018)).

formula suggested by McDonald and co-workers, that was also in part previously disclosed by Neto et al., that allows identifying the value of the Fermi velocity:

$$v = v_0 \frac{1 - 3\alpha'^2}{1 + 6\alpha'^2} \tag{2.2}$$

The v_0 is the Fermi velocity (which is usually 10^6 m/s) at the Dirac point and α' is the ratio between w and $\hbar v_0 k_\theta$ which is the interlayer dimensionless hopping amplitude. If α' is ≪1, in this case, the previous formula gives

$$v = v_0 \left(1 - 9\alpha'^2 \right) \tag{2.3}$$

In this specific case, the Fermi velocity goes to zero for $\alpha' = 1/\sqrt{3}$. In this case, if we took the previous formula:

$$\alpha' = \frac{w}{\hbar v_0 k_\theta} \tag{2.4}$$

where $k_\theta = G_K \theta$ and $G_K = 4\pi/\alpha$
and so for $\alpha' = 1/\sqrt{3}$:

$$\theta = \frac{w}{\hbar v_0 \alpha' G_K} = \sqrt{3} \frac{w}{\hbar v_0 \frac{4\pi}{3\alpha}} \tag{2.5}$$

This is the formula that allows to identify the first angle, which is around 1.1° after measuring the superconductivity dome (as we will explain in the next paragraph). Jarillo-Herrero and co-workers observed at this angle, for the first time, the flattening of the bands and the dramatic rise of the effective mass. The bandwidth observed was around 5–10 meV as shown in Fig. 2.7(c), where a logic enhancement of the DOS considers flattening.

Jarillo-Herrero and co-workers observed insulating states when the flat band where half-filled. As previously stated, this is the result of the competition between the quantum kinetic energy and of the Coulomb energy and leads to the definition of a state with characteristics coherent with a Mott-like insulator behaviour (as already explained). In order to move away from these insulating states, it is necessary to tune the charge density and so to avoid the half-filled situation. This can be easily done by changing the applied voltage exploiting a device configuration such as that shown in Fig. 2.7(a). In this case, we observe that the set-up is composed by full-encapsulated graphene monolayers, using boron nitride layers, originated by the same exfoliated flake. Researches from MIT were able to twist with a precision of around 0.1° and 0.2° and were able to approach the precision reached previously by Macdonald et al. in 2016. The graphene monolayers stacked were etched into a "Hall" bar and contacted from the edges (see Fig. 2.7a). As shown in Fig. 2.7(c) and

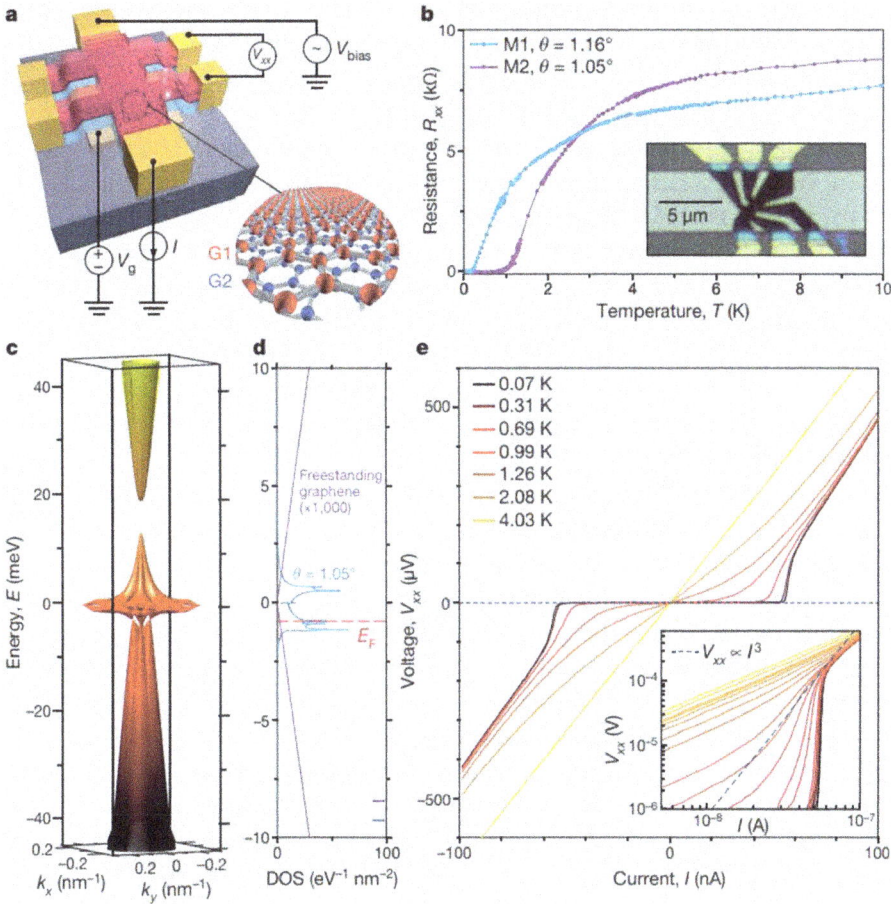

Fig. 2.7: Two-dimensional superconductivity highlighted in a graphene Moiré superlattice. (a) Schematic figure of a twisted bilayer graphene-based device and set-up for tests. (b) Four-probe resistance measure performed on two devices having twist angles of $\theta = 1.16°$ and $\theta = 1.05°$, respectively. (c) The band energy E of TBG at $\theta = 1.05°$ in the first mini-Brillouin zone of the superlattice. (d) The DOS corresponding to the bands shown in (c) ($\theta = 1.05°$). (e) Current–voltage curves for device M2 measured at $n = -1.44 \times 10^{12}$ cm^{-2} and various temperatures (reprinted with permission from Springer Nature Customer Service Centre GmbH, Springer Nature [31], Copyright Springer (2018)).

(d), the DOS of the flat bands is around three orders of magnitude larger than that of two uncoupled graphene layers. However, they observed that the peak of the DOS did not correspond to the density to half-fill the band and the presence of two flat bands up and down the neutrality point. The important feature is that as soon as the Fermi level is under the neutrality point, we can tune the superconductivity of stacked layers. Indeed, Jarrillo-Herrero's team did not observe any superconductivity for energy higher than 0 in the upper flat band. The most important point of the discovery, in opposite to previous studies on other superconductors, was that in this case it was necessary to apply a very low voltage that could induce a charge density of around 1.2×10^{12} cm^{-2} to move to the superconducting behaviour. This is the lowest density measured on superconductors.

Modulating the voltage applied, researchers were able to define a mapping of the conducting and insulating states as a function of the filling of the flat bands and of the carrier density. They observed (see Fig. 2.8a) that for an angle of 1.16°, that in case of ±2 or ±3 electrons for each superlattice unit cell, a Mott-like insulating state was identified and was related to the competition between the quantum kinetic energy and the coulomb interaction between charges that generates gap inside the bands. In case of full-filling of each superlattice cell (±4), which correspond to a density or charges of $\pm 3.2 \times 10^{12}$ cm^{-2}, the insulating states are related to the single-particle band gaps in the band structure. However, near the half-filling gap carrier density, some superconducting domes were clearly identified for both the angles (1.16° and 1.05°) that were very similar to what already observed in cuprates. Researchers from MIT identified a current of around 50 nA for a density of -1.44×10^{12} cm^{-2} and observed a clear sudden switch to a superconductivity trend. Moving to lower current, a metallic behaviour was recovered. The results obtained by MIT scientists were confirmed around 1 year later by McDonald et al. in 2019 [35]. In this paper, authors claimed to be able to build up a system to smoothly and precisely twist the angle between two layers of graphene stacked up. As in the previous study of MIT, for angle near 1.10°, they obtained insulator states for integer number of filling of the Moiré unit cell zones such as ±1, 2, 3, 4 (which correspond to the number of electrons: holes for each unit Moiré cell). McDonald and co-workers were able to calculate the activated band gap of the correlated insulating states exploiting the temperature/transport dependence on temperature up to 10 K which were, respectively, 0.34, 0.37 and 0.25 meV for −2, 2 and 3. They observed that the gap was largely lower for 1 and 0.14 meV, and absent for −3 and −1 states, which seems to confirm that these are mainly semimetallic state and not insulating ones. The main significance of the results obtained by MacDonald et al. in 2019 is that the superconducting domes position in the diagram was more delineated, with a shape defined more precisely, compared to the previous measures. They also observed new domes specifically between 0 and 1 and −1. These were and are the superconductivity states with the lowest DOS ($\sim 3 \times 10^{11}$) that have been observed up to now. This was due to the more homogeneity and reliability of the twist angles in the samples compared to those exploited by Jarillo-Herrero and co-workers who allowed pointing out the finer structure of materials. Moreover, McDonald and co-workers showed that, in case of the superconducting dome at −2 of Moiré band filling, the critical temperature was in this case 3 K which was quite larger than the 1.7 K found in the previous works. It is clear that this contribution was very interesting in identifying new superconducting states, thanks to a better systems allowing a finer twisting, but the most important role was to confirm the major breakthrough disclosed by experimental works of

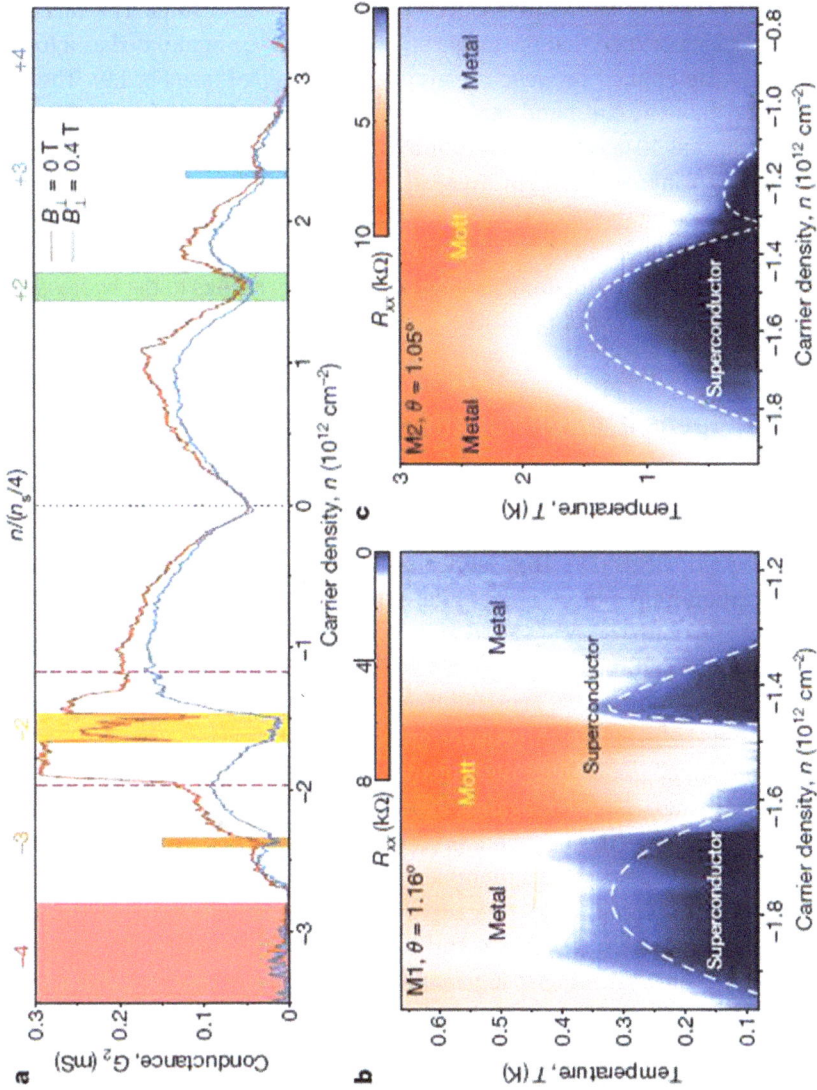

Fig. 2.8: Gate-tunable superconductivity in magic angle TBG. (a) Two-probe conductance $G2 = I/V$ bias of device M1 ($\theta = 1.16°$) measured in zero magnetic field (red) and at a perpendicular field of $B\perp = 0.4$ T (blue). (b) Four-probe resistance R_{xx}, measured at densities corresponding to the region bounded by pink dashed lines in (a) versus temperature. (c) Tests on device M2 (reprinted with permission from Springer Nature Customer Service Centre GmbH, Springer Nature [31], Copyright Springer (2018)).

Jarillo-Herrero's team at MIT. Other works confirmed these results. The Jarrillo-Herrero's team also outlined that the magic angle value can be modulated as a function of the pressure applied between the graphene layers [36] (see Fig.2.9). The results confirmed that the flat band regime reached at 5% compression for a twist angle of 1.47° and at 10% compression for a twist angle of 2.00°. This result was also confirmed by Yankowitz et al. in 2019 [37].

Yankovitz and co-workers [37] demonstrated that superconductivity could be enabled also in correspondence of twist angle larger than 1.1°, where correlated phases are otherwise absent, by modulating the interlayer spacing, thanks to the hydrostatic pressure and applying an out-of-plain strain (see Chapter 5 on straintronics).

2.5 What happens if we pile up three layers?

Jarrillo-Herrero and co-workers in 2021 decided to move a step forward and have recently started to test the superconductivity of three-layer graphene under the effect of very strong magnetic fields [38]. They fabricated a three-layer platform by stripping a thin layer of carbon from a block of graphite, stacking three layers, and rotating the central layer of 1.56° with respect to the outer layer. After that, they fabricated electrodes on the structure and they turned on a large magnet in the laboratory, directing the magnetic field parallel to the material. It was observed that when they increased the magnetic field around the three-layer graphene, it remained strong until the point before the superconductivity disappeared, as highlighted in Fig. 2.10 for two different doping values, but then, against expectations, reappeared at stronger magnetic field strengths. Indeed, to be more precise, the superconducting phase present at a parallel magnetic field = 0 is suppressed at a value of 8 T, and come back for larger values. This "coming back" to a superconducting behaviour is pointed out only in the region near zero resistance. This is extremely rare and usually does not occur with conventional spin singlet superconductors. They also observed that superconductivity lasted up to 10 T after "re-entry".

About 10 T is the maximum electric field strength value that can be achieved by a magnet in a laboratory and it is nearly three times what a superconductor should withstand when it is a conventional spin singlet, according to Pauli's limit, a theory predicting the maximum magnetic field allowing a material to demonstrate superconductivity. The reproduction of the superconductivity of a three-layer graphene, coupled with its persistence in higher magnetic fields than expected, does not permit to consider that the material as a common superconductor. A clear explanation of the physics behind the phenomenon has not been reached yet even if in the paper of Cao and co-workers, more hypothesis have been formulated but deeper investigations are necessary. Instead, it is probably a very rare type, probably a spin triplet, hosting a Cooper pair that passes through the material at high speeds unaffected by high magnetic fields. Jarrillo-Herrero's team plans to drill down the material

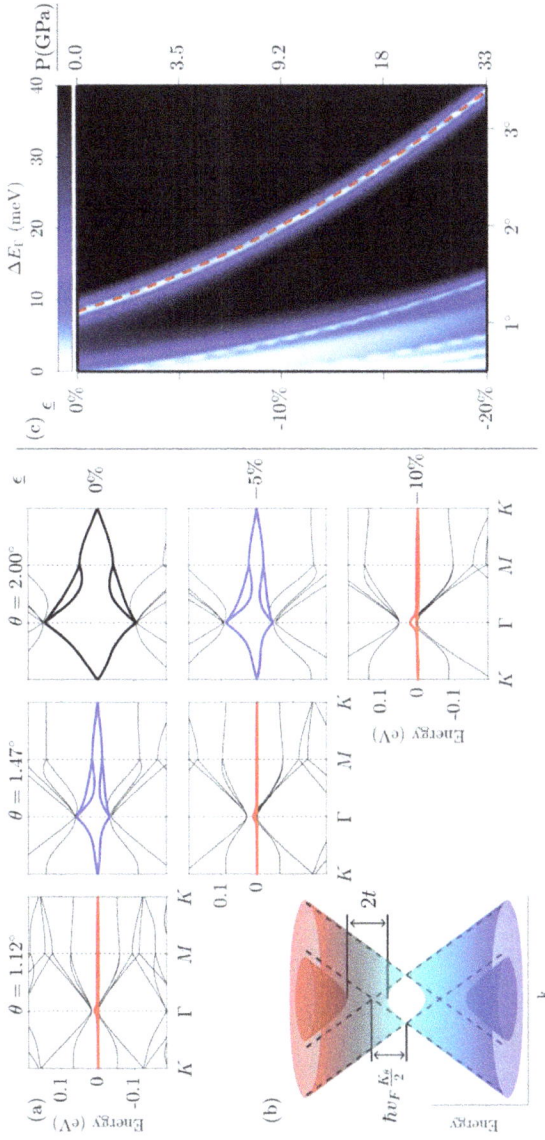

Fig. 2.9: (a) Band structures for twisted bilayer graphene under compression using the the ab initio tight binding model. (b) The shifting of the two coupled Dirac cones under pressure. (c) Critical values of the compression parameter ε (reprinted with permission from [36], Copyright American Physical Society (2018)).

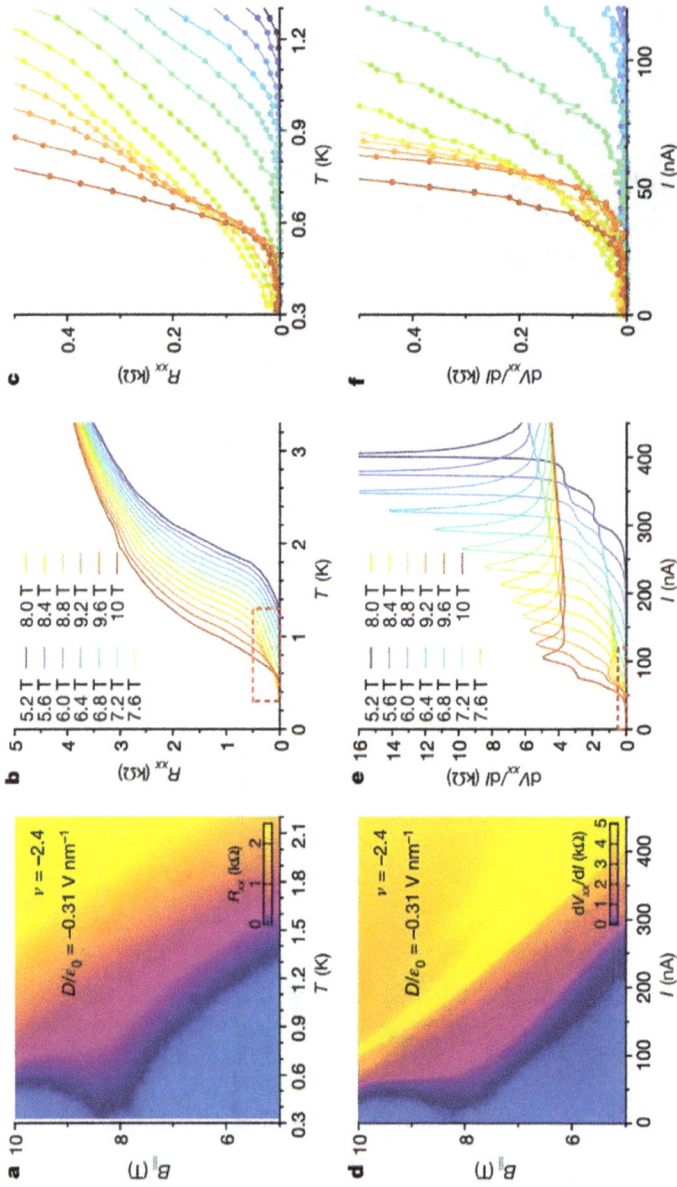

Fig. 2.10: (a) Resistance (R_{xx}) plotted as a function of the parallel magnetic field and of temperature. (b) and (c) Sections of R_{xx} plotted against T (b) and magnified view (c) at different values of $B\parallel$. (d) Differential resistance dV_{xx}/dI plotted as a function of the parallel magnetic field and I. The trend of I and temperature is similar. (e) and (f) Sections of dV_{xx}/dI plotted against I (e) and magnified view (f) at different values of the parallel magnetic field pointing out similar non-monotonic behaviour as in (b) and (c), strongly highlighting re-entrant superconducting phase (reprinted with permission from Springer Nature Customer Service Centre GmbH, Springer Nature [38], Copyright Springer (2021)).

to ensure accurate spin conditions. Indeed, this could help to define a design of more powerful magnetic resonance imaging machines and more robust quantum computers, even if we are very far to a suitable real application of these phenomena considering the extremely low temperatures employed to highlight them.

2.6 Conclusions and perspectives

Twistronics is only at its very beginning. The theory behind this new branch of condensed matter physics has not been exhaustively explored and understood yet. There are many issues to tackle such as the facility to implement these stacked layers and their exotic properties in devices that can be tested. The reproducibility remains a major problem. We have also discussed about the crystallographic quality of the monolayers considering that impurities can have a great influence on the final performances. Some works have also pointed out the importance of heterogeneous strain in the layers to justify the non-reproducible results between the different groups. Moreover, the temperatures needed to implement superconductivity are too low. The sacred Graal for superconductivity is to realize it at ambient temperature and we are very far from it. Strong efforts have to be performed to move from dream to reality but it is only the beginning. Maybe the solution will not be to use graphene layers but on other 2D materials that stacked will allow implementation of superconductivity at higher temperatures. However, we are not approaching this result, at the moment. Only by increasing the temperature for tests, we will be able to allow twistronics moving from being considered only as a sort of curiosity or a topic to produce high-rated papers to be a suitable strategy that can revolutionize science. Another rising approach is to stack more layers and to study the properties of these new structures. Many theoretical studies have been achieved on that. However, the scientific conclusions do not constitute a real step forward compared to the previous studies on bilayer graphene also considering the values of temperature and of magnetic field employed which is very large (10 T). In order to overcome the simple condition of "interesting object of research", more studies have to be performed on the bilayer structures to achieve a global overview of all the main issues to implement these structures in a real device or at less to understand how in a very long term, the properties of the magic angle could be exploited in devices. At the place to make a jump ahead, neglecting the studies on the simplest systems to find a sort of originality (e.g. adding more stacked layer), we need to perform more deeper studies on the bilayer system and especially on the test set-up allowing to achieve reproducible conclusions. Only after an exhaustive explanation of the physics behind this extremely simple system, we will be able to move forward and identify solutions, for example, to increase the temperature for achieving superconductivity. We have to come back to the approach followed by McDonald and co-workers who decided to study the simplest example to achieve conclusions on the larger one. To do that we have to give time to the research not only in optics to win a sort of race but to really understand the physics behind the magic angle phenomenon.

References

[1]   Carr, S., Massatt, D., Fang, S., Cazeaux, P., Luskin, M., Kaxiras, E. Twistronics: Manipulating the electronic properties of two-dimensional layered structures through their twist angle. Phys. Rev. B. 95, 2017, 075420, https://doi.org/10.1103/PhysRevB.95.075420.

[2]   Cao, Y., Fatemi, V., Fang, S. et al., Unconventional superconductivity in magic-angle graphene superlattices. Nature. 556, 2018, 43–50. https://doi.org/10.1038/nature26160.

[3]   Bistritzer, R., MacDonald, A. H. Moiré pattern in twisted double-layer graphene. PNAS. 108, 30, 2011, 12233–12237. https://doi.org/10.1073/pnas.1108174108.

[4]   Cao, Y., Fatemi, V., Demir, A. et al., Correlated insulator behaviour at half-filling in magic-angle graphene superlattices. Nature. 556, 2018, 80–84. https://doi.org/10.1038/nature26154.

[5]   Blundell, Stephen: Superconductivity: A Very Short Introduction. (Oxford University Press, 1st edition, 2009, p. 20), doi:10.1093/actrade/9780199540907.001.0001

[6]   https://www.nobelprize.org/prizes/physics/1913/onnes/biographical/

[7]   Mann, A. High-temperature superconductivity at 25: Still in suspense. Nature. 475, 2011, 280–282, https://doi.org/10.1038/475280a.

[8]   Meissner, W., Ochsenfeld, R. Ein neuer Effekt bei eintritt der Supraleitfähigkeit. Naturwissenschaften. 21, 1933, 787–788. http://dx.doi.org/10.1007/BF01504252.

[9]   Bardeen, J., Cooper, L. N., Schrieffer, J. R. Microscopic theory of superconductivity. Phys. Rev. 106, 1, April 1957, 162–164. https://doi.org/10.1103/PhysRev.106.162.

[10]   https://www.nobelprize.org/prizes/physics/1972/summary/

[11]   Kadin, A. M. Spatial structure of the Cooper pair. J. Supercond. Nov. Magn. 20, 2007, 285–292, https://doi.org/10.1007/s10948-006-0198-z.

[12]   Shigeji, F., Kei, I., Godoy, S. Quantum Theory of Conducting Matter. 2009, Springer Publishing, 15–27. ISBN978-0-387-88211-6https://link.springer.com/content/pdf/10.1007%2F978-0-387-88211-6.pdf.

[13]   https://dc.edu.au/wp-content/uploads/cooper-pair-phonon.png

[14]   Bednorz, J. G., Müller, K. A. Possible high Tc superconductivity in the Ba–La–Cu–O system. Z. Phys. B Condens. Matter. 64, 1986, 189–193, https://doi.org/10.1007/BF01303701.

[15]   Superconductivity at 93 K in a new mixed-phase Y-Ba-Cu-O compound system at ambient pressure. Wu, M. K., Ashburn, J. R., Torng, C. J., Hor, P. H., Meng, R. L., Gao, L., Huang, Z. J., Wang, Y. Q., Chu, C. W., Phys. Rev. Lett. 58, 1987, 908, https://doi.org/10.1103/PhysRevLett.58.908.

[16]   Greenwood, N. N., Earnshaw, A. Chemistry of the Elements. 2nd. 1997, Butterworth-Heinemann, ISBN 978-0-08-037941-8.

[17]   Monthoux, P., Balatsky, A. V., Pines, D. Toward a theory of high-temperature superconductivity in the antiferromagnetically correlated cuprate oxides. Phys. Rev. Lett. 67, 1991, 3448–3451, https://doi.org/10.1103/PhysRevLett.67.3448.

[18]   Lopes Dos Santos, J. M. B., Peres, N. M. R., Neto, A. H. C. Graphene bilayer with a twist: electronic structure. Phys. Rev. Lett. 99, 2007, 256802, https://doi.org/10.1103/PhysRevLett.99.256802.

[19]   Kim, K., DaSilva, A., Huang, S., Fallahazad, B., Larentis, S., Taniguchi, T., Watanabe, K., LeRoy, B. J., MacDonald, A. H., Tutuc, E. Tunable moiré bands and strong correlations in

small-twist-angle bilayer graphene. PNAS. 114, 13, 2017, 3364–3369. https://doi.org/10.1073/pnas.1620140114.

[20] Dean, C., Young, A., Meric, I. et al. Boron nitride substrates for high-quality graphene electronics. Nat. Nanotech. 5, 2010, 722–726. https://doi.org/10.1038/nnano.2010.172.

[21] Wang, L. et al. One-dimensional electrical contact to a two-dimensional material. Science. 342, 6158, 2013, 614–617. 10.1126/science.1244358.

[22] Zhang, Y., Tang, T. T., Girit, C. et al. Direct observation of a widely tunable bandgap in bilayer graphene. Nature. 459, 2009, 820–823. https://doi.org/10.1038/nature08105.

[23] Weitz, R. T., Allen, M. T., Feldman, B. E., Martin, J., Yacoby, A. Broken-symmetry states in doubly gated suspended bilayer graphene. Science. 330, 6005, 2010, 812–816. 10.1126/science.1194988.

[24] Lee, K., Fallahazad, B., Xue, J., Dillen, D. C., Kim, K., Taniguchi, T., Watanabe, K., Tutuc, E. Chemical potential and quantum Hall ferromagnetism in bilayer graphene. Science. 345, 6192, 2014, 58–61. 10.1126/science.1251003.

[25] Maher, P., Wang, L., Gao, Y., Forsythe, C., Taniguchi, T., Watanabe, K., Abanin, D. et al. Tunable fractional quantum Hall phases in bilayer graphene. Science. 345, 6192 July 3, 61–64. 10.1126/science.1252875.

[26] Sanchez-Yamagishi, J. D. et al. Quantum Hall effect, screening, and layer polarized insulating states in twisted bilayer graphene. Phys. Rev. Lett. 108, 7, 2012, 076601. https://doi.org/10.1103/PhysRevLett.108.076601.

[27] Ponomarenko, L., Gorbachev, R., Yu, G. et al., Cloning of Dirac fermions in graphene superlattices. Nature. 497, 2013, 594–597. https://doi.org/10.1038/nature12187.

[28] Dean, C., Wang, L., Maher, P. et al., Hofstadter's butterfly and the fractal quantum Hall effect in moiré superlattices. Nature. 497, 2013, 598–602. https://doi.org/10.1038/nature12186.

[29] Hunt, B. et al., Massive Dirac fermions and Hofstadter butterfly in a van der Waals heterostructure. Science. 340, 6139, 2013, 1427–1430. 10.1126/science.1237240.

[30] https://www.quantamagazine.org/how-twisted-graphene-became-the-big-thing-in-physics-20190430/?mc_cid=b706c722b6&mc_eid=a2bccb49a5

[31] Cao, Y., Fatemi, V., Fang, S. et al., Unconventional superconductivity in magic-angle graphene superlattices. Nature. 556, 2018, 43–50. https://doi.org/10.1038/nature26160.

[32] Cao, Y., Fatemi, V., Demir, A. et al., Correlated insulator behaviour at half-filling in magic-angle graphene superlattices. Nature. 556, 2018, 80–84. https://doi.org/10.1038/nature26154.

[33] Mott, N. F., Peierls, R. 1937 Discussion of the paper by de Boer and Verwey. Proceedings of the Physical Society. 49 (4S): 72, https://doi.org/10.1088/0959-5309/49/4S/308

[34] Wang, Z., Liu, C., Liu, Y., Wang, J. High-temperature superconductivity in one-unit-cell FeSe films. J. Phys. Condens. Matter. 29, 2017, 153001, https://doi.org/10.1088/1361-648x/aa5f26.

[35] Lu, x., Stepanov, P., Yang, W. et al., Superconductors, orbital magnets and correlated states in magic-angle bilayer graphene. Nature. 574, 2019, 653–657. https://doi.org/10.1038/s41586-019-1695-0.

[36] Carr, S., Fang, S., Jarillo-Herrero, P., Kaxiras, E. Pressure dependence of the magic twist angle in graphene superlattices. Phys. Rev. B. 98, 2018, 085144, https://doi.org/10.1103/PhysRevB.98.085144.

[37] Yankowitz, M., Chen, S., Polshyn, H., Zhang, Y., Taniguchi David Graf, W., Young, A. F., Dean, C. R. Tuning superconductivity in twisted bilayer graphene. Science. 363, 6431 2019, 1059–1064. 10.1126/science.aav1910.

[38] Cao, Y., Park, J. M., Watanabe, K. et al., Pauli-limit violation and re-entrant superconductivity in moiré graphene. Nature. 595, 2021, 526–531. https://doi.org/10.1038/s41586-021-03685-y.

3 Valleytronics and a new way to encode information using 2D materials

3.1 Valleytronics: basic science behind and interest

In crystalline systems, we can determine the relationship between the crystal momentum and the energy associated with an electron through the electronic band. We commonly observe in intrinsic semiconductors that the Fermi energy usually lies between the highest valence band and below the lowest conduction band. The highest point and the lowest point in terms of energy of, respectively, the conduction and valence bands are defined as crystal momentum space "valleys". These energy levels can be occupied by electrons. As a consequence, we can imagine to attribute to a single particle as intrinsic properties, its charge, its spin and definitively its belonging to one specific valley (which can be defined as a band). This definitively corresponds to a new freedom degree that can be attributed to a particle (electron). With the term "valleytronics" we define the branch of studies that deals with the manipulation of this new degree of freedom to store more information (charge, spin and valley), exploiting mainly optical stimuli to excite electrons. This is an extremely interesting approach considering that can lead to the manipulation and storage of information, reducing in a dramatic way the energy consumption compared to charge storage devices exploiting charge (e.g. flash memories [1]). Moreover, it can open the way for the storage of information through optical stimulation of charges using specific wavelength, increasing dramatically the amount of information and at the same time the time for storage. A future field of applications can be in the framework of electronic devices (especially in opto-electronics) and of quantum computing technology. Indeed, thanks to valleytronics we can surpass theoretically existing quantum information storage system based on charge or spin control technology [2] (e.g. hard disk). Scientists are leading research on valley control since valleytronics has a huge potential that brings together spintronics and nanoelectronics in the next generation of devices also thinking in a beyond CMOS perspective. However, all these advantages stay only theoretical because the practical realization of valleytronics has not achieved yet considering the evident technological issues that do not allow securing the stability and enough quantity of valleys to implement valleytronics in real devices. The objective of this chapter is to explain step by step why a material can be fit for valleytronics and how information can be practically stored.

https://doi.org/10.1515/9783110656336-004

3.2 Which materials for valleytronics?

Valleytronics is not exclusively related to 2D material properties. Indeed, it has been studied for a quite long time. The main material characteristics expected to implement valleytronics have a material owning a band structure composed of two (or more) degenerate but inequivalent valley states (local energy extrema). The last needs to be handled to write, process and store information. The first studies on the topics were performed more than 10 years ago [3]. Indeed, one of the common approaches used was to employ strain to modulate the energies of valleys (creating degeneracy) as already done in the semiconductor electronics industry to increase carrier mobilities [4]. In more fundamental research, some traditional semiconductor properties have been studied, for example in silicon and aluminium arsenide (AlA). These last own multiple valleys in the conduction bands located at or near the X symmetry points in the Brillouin zone. Concerning silicon, we can mention older works back to the 1970s, with scientific works on inversion layers occurring at silicon/insulator interfaces. For example, Sham et al. [5] suggested a different interpretation of what is observed in a silicon metal–oxide–semiconductor field-effect transistor with an interface of Si tilted away from the high-symmetry plane (001) to a high-index plane such as (119), suggesting for the first time the existence of degenerate valleys. Actually, Cole, Lakhani and Stiles [6] observed structures in the dc conductivity and anomalous oscillations in the magnetoconductivity. To understand this phenomenon, they proposed a model of a 1D superlattice along the interface. However, the precise cause of the formation of the superlattice was not explained. Sham and co-workers suggested that this phenomenon was related to a splitting between the two band minima of the conduction band. It is important to underline that the phenomenon was not observed in 2D materials but in bulk ones. Fusayoshi et al. [7] had already tried to explain the existence of a sort of valley splitting in the same system. Indeed, they outlined that to observe in a better way the splitting between the band was necessary to apply a magnetic field. For this reason, after two decades, strain and magnetic field were finally employed to handle and change valley polarization in 2D electron–gas systems hosted in AlA heterostructures [8]. In this case, the splitting of the conduction band valleys in high-mobility 2D electrons was confined to AlA quantum wells. The results pointed out that while the valleys were nearly degenerate in the absence of magnetic field, when a perpendicular magnetic field is applied, they split. The splitting was mainly related to the strength of the perpendicular part of the magnetic field that seemed to highlight that its origin could be due to an electron–electron interaction. Recently, some new impressing advances have been performed in handling correctly valley-polarized currents in diamond [9], giving rise to valley polarization in bismuth, and allowing the determination of the energy difference between nearly degenerate valleys in quantum dot and donor-bound silicon systems [10–13]. These experiences have shown that many crystalline systems can feature electronic valleys that can be exploited to implement valleytronics and so where the polarization can be obtained, thanks to the valley-dependent energy dispersion of carriers. The difficulty to exploit this phenomenon in a suitable way to move forward potential applications remains the main drawback to lead to the implementation of valleytronics in real devices. Indeed, up to now the main issue was that materials could not be used to implement valleytronics without applying magnetic fields or strain or other external solicitations. We can say that the valleytronics is not an intrinsic property of these materials. The ability to exploit valley polarizations could not be achieved in an effective way exploiting common materials. This changed with the emergence of 2D materials showing honeycomb structure, as in case of graphene. What is the reason? The impressing advantage of using 2D materials is that they show degenerated bands without applying an external stimulus to "create" them. They intrinsically own the typical characteristics of materials fit for achieving valleytronic studies. For this reason, the advent of 2D materials constitutes a real revolution for this specific field of research. The growing focus of scientific works on valleytronics of 2D material is mainly related to the graphene-like band structure featuring two degenerate valleys at $+K$ and $-K$ (also referred to as K and K_0) points in the crystal momentum space (see Fig. 3.1). A quite large panel of 2D materials such as graphene, hexagonal boron nitride (hBN) and transition metal dichalcogenides (TMDs) shows these characteristics. Now we will try to explain in a simple way why in hexagonal 2D materials, such as graphene and monolayer group VI TMDs (e.g., MoS_2, $MoSe_2$, WS_2 and WSe_2) [14], the electronic properties at the band edge are strongly influenced by the two inequivalent valleys at the $+K$ and $-K$ points at the edges of the Brillouin zone (see Fig. 3.1a). It should be noted that in case of intrinsic graphene, the last one shows inversion symmetry because graphene has a hexagonal network of identical carbon atoms. The main consequence is that the K valleys cannot be differentiated. Xiao and co-workers [15] demonstrated that, breaking the inversion symmetry applying an external field or exploiting interactions with substrate, it was possible to differentiate valleys, allowing bands showing valleytronic-related features. However, in case of monolayer TMDs, this naturally lacks inversion symmetry (considering that it is composed of different atoms in a hexagonal network) and, therefore, owns valley-contrasting physical effects [16]. In this section, maybe the most important valleytronic effects, also in terms of potential applications, is reviewed: valley-dependent optical selection rule of monolayer TMDs.

In TMDs, valleys can be modelled using a binary pseudo-spin behaving like a spin-1/2 system; the electrons in the $+K$ valley are defined as valley pseudo-spin-up, and the electrons in the $-K$ valley as valley pseudo-spin-down. To be clearer, for pseudo-spin we mean a property of the particle that can be assimilated to an intrinsic property such as the spin. Therefore, in a doped system, a carrier population distribution polarized in a $+K$ or $-K$ valley can store binary information (using the degenerated valleys) and this information will be characterized for each particle by the charge, the spin and the valley spin or valley pseudo-spin (three information for only one particle). In systems that do not show inversion symmetry, carriers in $+K$ and $-K$ valleys undergo opposite Berry curvatures, acting as effective magnetic fields in the momentum space,

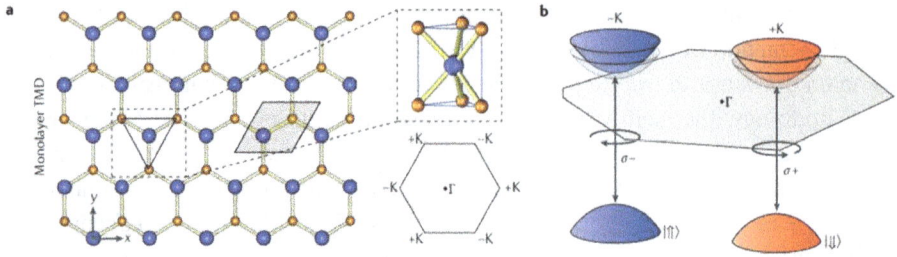

Fig. 3.1: (a) Two-dimensional hexagonal structure of a monolayer TMD material. In blue, transition metal atoms; and in orange, chalcogen atoms. Hexagonal Brillouin zone labelling Γ and +K, −K, the two inequivalent points considering the broken symmetry of the lattice (not like graphene case). (b) Valley polarization-dependent rules [17] (reprinted with permission from Springer Nature Customer Service Centre GmbH, Springer Nature [17], Copyright Springer (2016)).

and the electronic equations of motion are regulated by semiclassical transport equation. The Berry curvature, as previously highlighted in Chapter 1, allows detailing the geometric properties of the electronic bands. It is a key concept that permits completely to grab the physical meaning of band topology-related effects. If we adopt a semiclassical way to describe electronic motion in crystal lattices, the electrons can be assimilated to a Bloch wave flowing through the crystal lattice. Its mean velocity is proportional to the gradient of the electronic energy of the band. In this schematization, the Bloch's form of electron wave function bears consideration of the lattice periodicity and of the reaction of carriers undergoing the effect of a magnetic or electric field. Moreover, strictly related to the Berry curvature of the electronic band, an anomalous speed emerges which is transverse to the applied electric field. Indeed, this speed is a fundamental parameter implementing valley current and associated effects in materials that feature a non-vanishing Berry curvature.

A magnetic momentum related to the valley magnetic emerges from the orbital angular momentum, which is strictly connected to the Berry-phase effects on electrons due to the band structure [18, 19]. We observe that in systems showing strong SOC, the time-reversal symmetry conducts the spin splitting having opposite signs at the +K and −K valleys, leading to an effective coupling between spin and valley pseudo-spin [20]. As a consequence, strong spin–valley coupling can have an interesting effect such as the increase of the spin and valley polarization lifetime [21], the handling of the spin by exploiting the valley properties, and finally, the potential to pair them to layer pseudo-spin (and electrical polarization) in multilayers with a specific stacking order [22–25]. In the next paragraph, we will discuss about the way to manipulate the pseudo-spin in TMD materials through optical stimulus allowing encoding information in valleys.

3.3 Optical stimulus: the right way to encode valley pseudo-spin?

In this section, we will mainly focus our analysis on monolayer group VI TMD semi-conductors, as already evoked, having the chemical formula MX_2, where M = Mo, W, and X = S, Se or Te, which show interesting features in the framework of valley-tronics. It should be noted that TMD semiconductors received considerable experi-mental investigations decades ago [26–28] and were rediscovered in the context of 2D van der Waals materials following graphene. One of the main characteristics that captured the attention of scientists, as in case of black phosphorous (see Chapter 4), concerned its optical properties and their strongly layer-dependent properties. This specific feature was highlighted through photoluminescence (PL) measurements [29, 30]. To describe the test set-up, in PL spectroscopy, focused light (commonly using a laser) gives rise to the excitation of electrons and holes in a material. Then, these par-ticles are subject to relaxation processes and then to recombination with each other with consequent photon emission. Usually, the intensity of the PL is analysed as a function of emission energy (or wavelength) in order to produce a PL spectrum as shown in Fig. 3.2 for MoS_2. What is peculiar in case of MoS_2 is that a layer is a direct band gap semiconductor that can emit PL in a suitable way. However, when we have to deal with a bilayer MoS_2, in this case, scientists observed that the interlayer cou-pling transforms the band structure (lowering the gamma point valence band energy) and a strong reduction of the PL signal, considering that the gap became indirect.

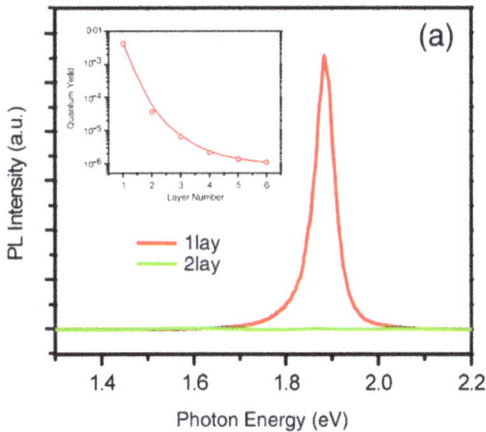

Fig. 3.2: Layer-dependent PL from MoS_2 [31] (reprinted figure with permission from [31], Copyright (2010) by the American Physical Society).

It has to be pointed out also that composite quasi-particles, such as excitons, own a valley degree of freedom because of the localization of the associated electrons and holes in the $+K$ and $-K$ valleys. An exciton is defined as a neutral quasi-particle and is the result of a bound state achieved between an electron and a hole that undergo an electrostatic Coulomb force. This quasi-particle appears in insulators or semiconductors. The same thing for trions which are similar to excitons but are the result of the interaction of three particles (e.g. two electrons and a hole). The peculiarity of excitons is that they can be associated with an elementary excitation in solid-state matter and that can only transport energy but does not bring electrical charge [32, 33].

The remarkable characteristic of these excitons is that they host a binary valley pseudo-spin that can be optically addressed (i.e. excited and read-out) with circularly polarized light. As a function of the polarized light, we will be able to "store" precise information in the form of valley pseudo-spin. This is possible, thanks to the broken inversion symmetry of TMD systems, explained in the previous sections, that gives rise to a valley-dependent optical selection rule [34, 35]. Thanks to that, it can be selected and targeted in an interband transition using a specific circularly polarized light: a right circularly polarized light leads to transitions in the $+K$ valley, and left to transitions in the $-K$ valley (see Fig. 3.3). Thanks to that we can store information as shown in Fig. 3.3.

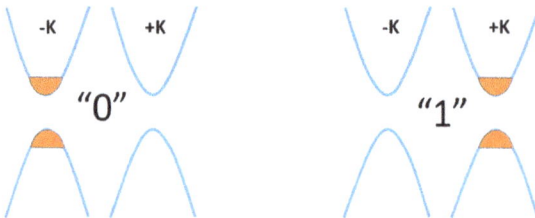

Fig. 3.3: Schematization about how information can be encoded through valleytronics as a function of the polarization of the employed polarized light.

Indeed, the first paper that highlights the selection rule was published in 2008 by Yao et al. [36] but the most surprising characteristics of this paper is that it focused on graphene. The immediate question is why graphene can highlight this kind of features considering that its inversion symmetry is preserved? In fact, scientists exploited the fact that when graphene was grown on hBN or on SiC, it could be covalently bonded to the substrate. Only in this way, the breaking of the inversion symmetry valleytronics could take place. In case of hBN, as highlighted by the pioneering theoretical work of Giovannetti et al. [37] in 2007, this wide gap insulator owned a layered structure that was very similar to that of graphene. However, the two atoms in the unit cell were chemically inequivalent. Therefore, placed on top of hBN the two carbon sublattices of graphene were no more identical, from a chemical point of view. This situation led to an interaction with the substrate that broke the inversion symmetry. The band structure calculations in the local density approximation demonstrated the evidence of a gap of at least 53 meV, an energy roughly twice as large as $k_b T$ at room temperature was, therefore, induced. In another work by Zhou et al. in 2007 [38], researchers demonstrated that if graphene was epitaxially grown on SiC substrate, it could be observed the emerging of a gap of ≈0.26 eV. This gap value changed in an inversely proportional way as a function of the sample thickness and finally went to zero if the number of layers exceeds 4. As in the previous case, the explanation for the creation of this gap was the breaking of sublattice symmetry owing to the graphene–substrate interaction. Coming back to the optical selection rule and to TMD materials, this effect has been shown through several optical experiments on monolayer. In these cases, the circularly polarized light was used to inject excitons into one specific valley, thereby emerging as a difference in the valleys' population. This valley polarization could be quite easily experimentally read out by measuring the circularly polarized components of the emitted PL intensity. The first impressing work that demonstrated this technique was published in 2012 by Mak et al. [39]. Scientists identified the MoS_2 as the perfect materials to highlight the valley polarization because it consists of a single layer of molybdenum atoms sandwiched between two layers of sulphur atoms in a trigonal prismatic structure (Fig. 3.3a).

At the K and K' valleys in momentum space, the extremities of the two bands are mainly of molybdenum d-orbital character. Considering that the inversion symmetry is intrinsically broken in this material, spin–orbit interactions split the valence bands by 160 meV [40–42] (Fig. 3.3b). The spin projection along the c-axis of the crystal, S_z, is well defined and the two bands are of spin-down ($E\downarrow$) and spin-up ($E\uparrow$) in character. This broken spin degeneracy, in combination with time-reversal symmetry is not altered ($E\downarrow(k) = E\uparrow(-k)$, where k is the crystal momentum), implies that the valley and spin of the valence bands are inherently coupled in monolayer MoS_2. Consequently, interband transitions at the two valleys are allowed for optical excitation of opposite helicity incident along the c-axis, that is, left circularly polarized ($s2$) and right circularly polarized (sþ) at the K and K' valleys, respectively (Fig. 3.1b). Scientists could demonstrate that it was possible to implement an optical selection rule of the valleys, through measurements performed exploiting low-temperature polarization-dependent PL [19]. Thanks to this technique, the excitons were optically excited using a circularly polarized light and then the material, TMD in this case, was studied using polarization optics before its detection. By rotating the polarization optics, it was possible to obtain the copolarized PL(co) and cross-polarized PL(cross). The valley polarization, η, gives us a measure of the contrast between co- and cross-polarized components and can be obtained using this formula:

$$PL = (PL(co) - PL(cross))/(PL(co) + PL(cross))$$

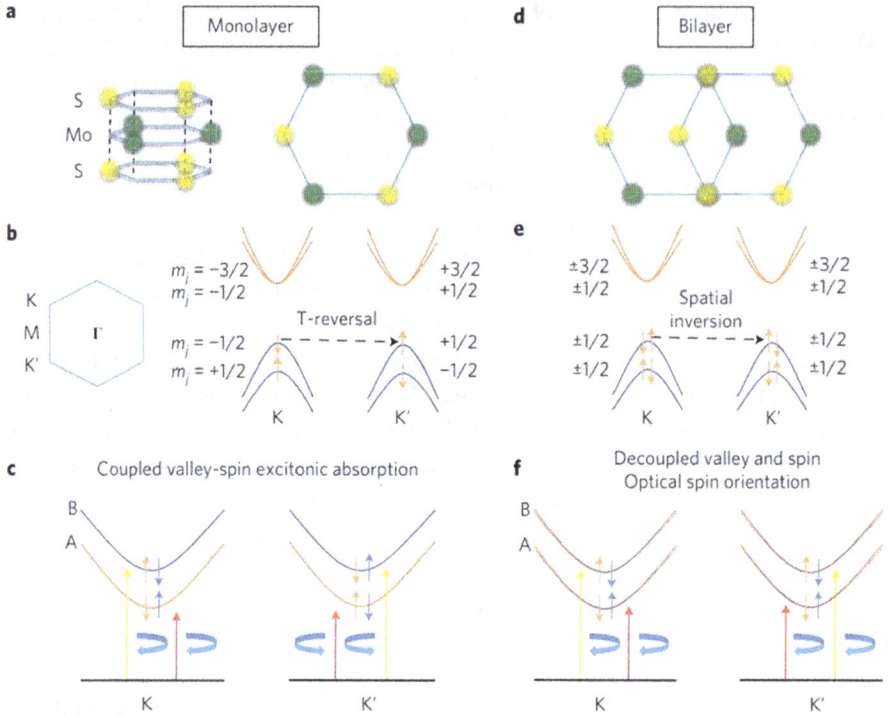

Fig. 3.4: Atomic structure and electronic structure at the K and K' valleys of monolayer (a–c) and bilayer (d–f) MoS_2. (a) The honeycomb lattice structure of monolayer MoS_2 pointing out that spatial inversion symmetry is not respected. (b) The lowest energy conduction bands and the highest energy valence bands labelled by the z-component of their total angular momentum. (c) Optical selection rules for the A and B exciton states. (d) Bilayer MoS_2 with Bernal stacking. (e) Spin degeneracy of the valence bands is restored by spatial inversion and time-reversal symmetries. Valley and spin are decoupled. (f) Optical absorption in bilayer MoS_2 (reprinted with permission from Springer Nature Customer Service Centre GmbH, Springer Nature [39], Copyright Springer (2012)).

The first measurements were carried out on monolayer MoS_2 [43], which showed large (50–100%) values of valley polarization for near-resonance excitation (see Fig. 3.4b). Then, in a similar way, measurements on monolayer WSe_2 pointed out that when the system was pumped with linearly polarized light, a coherent superposition of valley excitons could be generated and read out optically [44]. One of the interesting questions that could be raised is, why are the optical properties of monolayer TMDs dominated by excitons (Coulomb-bound electron–hole pairs) with large 500 meV binding energies? As demonstrated by Chernikov et al. [45], this large binding energy is mainly related to the common reduced screening of Coulomb interaction in 2D materials, in contrast to III–V semiconductors, which have exciton binding energies on the order of 5–10 meV.

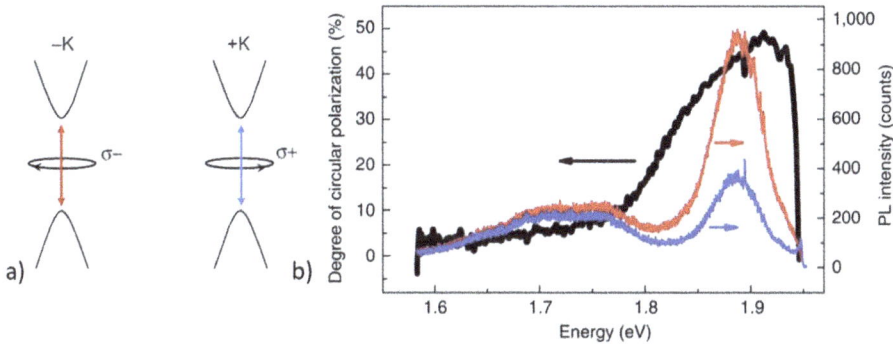

Fig. 3.5: (a) Valley-dependent optical selection rule for monolayer TMDs. (b) Circularly polarized micro-PL of monolayer MoS_2 at 83 K, along with the degree of circular polarization of the PL spectra. The red and blue curves correspond to the intensities of σ+ and σ− polarizations, respectively, in the luminescence spectrum. The black curve is the net degree of polarization (reprinted with permission from Springer Nature Customer Service Centre GmbH, Springer Nature [47], Copyright Springer (2012)).

In an extremely interesting way, PL measurements pointed out the existence of excitonic valley polarizations in monolayer TMDs of 50–100% see Fig.3.5. This result showed that it was possible to address in a precise and deterministic way the valley through optical stimuli [46–48]. As a consequence, a great interest is in how to handle these valley pseudo-spins to achieve practical valleytronic devices because only in this way we will be able to move forward the field of applications. The dream is to fabricate valleytronic circuits based on optically injected valley excitons where information could be stored exploiting the exciton valley pseudo-spin. However, the relatively short lifetime of excitons is a major hurdle to overcome, and valley depolarization times could strongly limit their practical implementation for valleytronics.

3.4 Potential implementation of valleytronics: what are the main challenges?

Different mechanisms can influence the valley lifetime. One of the major factors is the presence of scattering of impurities/defects. Reducing the density can increase in a dramatic way the valley lifetime. Other parameters that have not been deeply studied yet are the effect of phonons, disorder and nuclear spins. Figure 3.6 shows the state of the art concerning valley lifetime of electrons, holes, excitons and trions (excitons with an extra electron or hole), across several different TMDs as measured by circularly polarized PL and Kerr rotation techniques by various research groups. We have to point out that most of the data reported have been obtained through very temperature measurements (≈4 K) [55]. An extremely important criterion to judge the potential implementation of the different particles or quasi-particles in valleytronics is their lifetime and recombination times. These two need to be as large as possible to imagine their potential utilization to encode information. From this point of view, analysing Fig. 3.6, we can observe that electrons and holes seem to be more adapted compared to excitons and trions. However, the conclusion is not so straightforward. The other very important parameter is the valley coherence time. It is necessary to have a large coherence time if we want to keep the information unaltered. In case of electrons and holes, they are subject to perturbations related to the fact that the crystal momentum is variable. This factor reduced in a strong way their valley coherence, which is a major drawback. However, up to now, no measures of valley coherence have been performed, which strongly limit our analysis. It seems that excitons can show large valley coherence simply because they do not bring charge as they are quasi-particles, hence, they are not subject to interferences related to the proximity with other charges and also because they show one tightly bound state corresponding to each in the momentum space. It is evident that we are only at the beginning of these studies and that it is very difficult without reproducible and systematic tests to draw a conclusion and to evaluate all the pros and cons for each particle and quasi-particle.

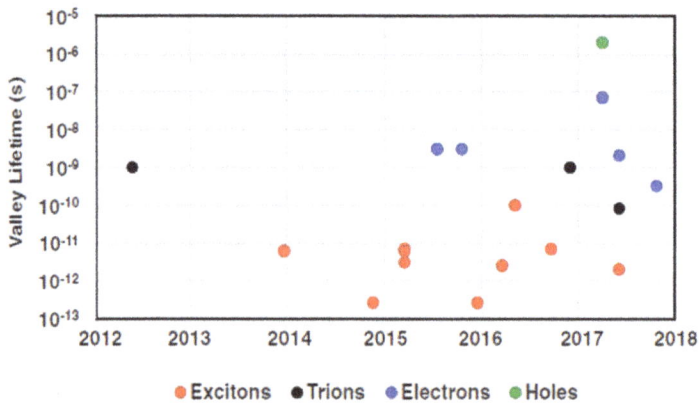

Fig. 3.6: Valley polarization lifetime coming from different studies for particles and quasi-particles in several TMD single layer (reprinted with permission from John Wiley and Sons Inc, Small [50], Copyright John Wiley and Sons Inc. (2018)).

As previously explained (see Fig. 3.3), the valley-selective excitons can allow to encode information in the next-generation optoelectronic and information processing devices. The main concern is that excitons in TMD monolayers tend to be short-lived and rapidly jump between the valleys, causing the stored information to be either quickly deleted or scrambled. To tackle this issue, it is necessary to develop strategies that allow realizing long-lived valley-selective excitons. Indeed, in case of aligned bilayers of TMDs, interlayer excitons have long lifetimes but suffer from rapid valley mixing. For this reason, Scuri et al. [45] fabricated WSe_2 bilayers with a twist angle between the layers (see Fig.3.7). To fabricate a suitable device, scientists encapsulated the samples in hBN. Thanks to that, they were able to prevent contamination and set-up a top and bottom graphene gate to enable independent control of doping and vertical electric fields. However, the valley coherence time, a crucial quantity for valley pseudo-spin manipulation, was and is very difficult to directly probe.

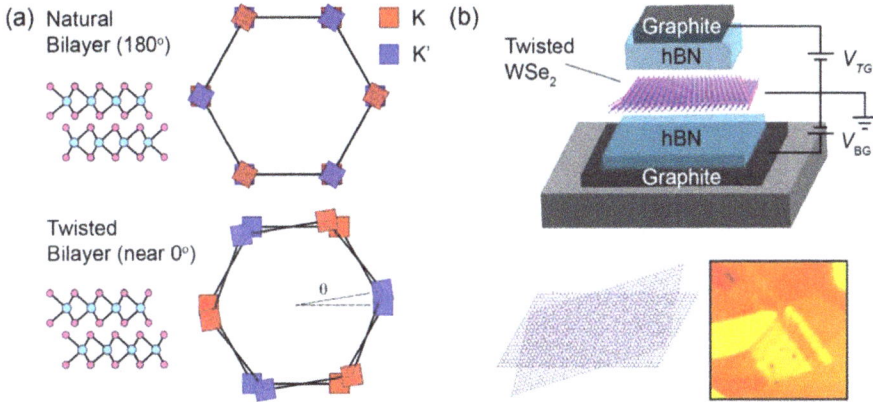

Fig. 3.7: Band engineering through twisting: (a) side view and Brillouin zone alignment of natural (2H-stacked, top) and twisted homobilayer (bottom). (b) Device schematic (top), illustration of a Moiré pattern in a twisted WSe_2/WSe_2 bilayer (bottom left) and optical image of a device (bottom right) (reprinted figure with permission from [45], Copyright (2020) by the American Physical Society).

The main results of the tests performed by Scuri and co-workers were that the interlayer excitons in twisted WSe_2/WSe_2 bilayers exhibited a high (>60%) degree of circular polarization and long valley lifetimes (>40 ns) at zero electric and magnetic fields. Thanks to this approach, the valley lifetime could be tuned by more than three orders of magnitude via electrostatic doping. A key effect of the twisting is that it changes the alignment between the valleys in two layers in the momentum space so that it is difficult for interlayer excitons in a given valley to hop to a different valley. The new findings open up new opportunities for tuneable chiral photonics and electrically switchable valleytronic devices. This is a major result that opens a new field of innovation for valleytronics to tackle real-life applications in a medium–long-term optics. Scientists have also suggested to incorporate the system in a cavity to extend the exciton lifetime and they are exploring twisted TMD bilayers as a platform for creating arrays of valley-selective quantum emitters, which could be used to realize solid-state quantum simulators.

3.5 Some examples of applications and perspectives

The focus of valleytronics research community logically up to now has mainly focused on material growth, characterization and valley physics experimentation, considering that we are talking about a quite new discovery. Presently, it is very difficult to harness valleytronics, excitons behaviour, to imagine the fabrication of devices exploiting it. We can anyway imagine some potential long-term applications that could be implemented considering the last developments. One of the major fields of application, which is natural considering the origin of the phenomenon, is the microphotonic elements, as suggested by Vitale et al. [49] in 2018. The interaction between light and matter that happens in TMDs with the possibility to address a specific energetic state strongly coupled with spin using the polarization of an incident light constitutes an extremely interesting approach for applications such as microscale nonreciprocal optics, circularly polarized light emitters or polarization-sensitive detectors. The most straightforward application is in the field of detectors. Indeed, thanks to 2D TMD, and using valleytronics we can imagine a fabricating camera that is able to detect the specific polarization of light. In this case, simply a thin film of 2D TMD is able to implement this kind of sensing and could help push further miniaturization of the system. Exploring the potential applications in quantum communication, quantum key distribution (QKD) [50] is a logical application considering that, in this case, information is usually encoded on single photons. In this protocol, it is necessary to create and detect photon with a specific polarization. Indeed, thanks to valleytronics and the intrinsic properties of TMDs such as MoS_2, which can be easily implemented with the possibility to achieve miniaturized optical systems at the microchip level avoiding the integration of encumbering optical elements. The QKD, thanks to this potential miniaturization brought by the utilization of TMDs, can be integrated to secure communication in mobile phone, tablets and also the information about personal health care. The interest in QKD technologies is booming, and in 2019, a European testbed has been created allowing the spreading of the technological advantages of this solution strongly affecting industrial applications in the field. TMDs and valleytronics can play a key role in this, potentially being used for emitters and detectors. However, one of the most impressing advance is the possibility to implement a new paradigm in quantum computing where the information is not stored, making flowing and modulation of charges through transistors, but exploiting the belonging of particles (electrons or holes) or quasi-particles (as in the case of excitons or trions) to one specific valley, exploiting the optical stimulation of the material. In the quantum computer developed by IBM and Google, the main issue concerns the operating temperature, which is very low around 80°K. Moreover, the energy scale for operations in case of a memory-based valleytronics concept will be governed by the splitting induced between the degenerated valleys at the bottom of the bands, which is of the order of 30 meV and so around 30 times lower than the commercial silicon CMOS technology. Since the device switching energy (in conventional transistors) scales quadratically with voltage, this would lead to a theoretical reduction of a factor 1,000 in the required power. This is a strong case for future valley-tronic-based computing. Moreover, in case of exploiting charge to store information, the problem is that this last can be perturbed by other charges or by temperature. In case of TMD, all materials own the valleytronic properties, and the pseudo-spin related to a valley is a robust property at ambient temperature that can be changed only after a strong intervention on the properties of the material [51–53]. Moreover, 2D TMD implements the ultra-thin film memory technology with the strong potential for 3D integration and so enhances the storage capacity in a huge way, which exploits an optically driven concept reducing dramatically the energy consumption compared to the existing technology. Some works in this direction have been performed in the last years but the valley state lifetime is an issue as previously explained. However, maybe the most important issue concerns the quality of the materials that have to be grown to show valleytronic features. Indeed, it is nearly impossible with the present technologies to perform the growth of large-scale mono-crystalline samples. The consequence is that polycristalline samples dont not allow highlighting the valleytronic features necessary to implement real devices because it is not possible to perform systematic tests on the materials and on the coherence time of electrons and holes. Considering that presently it is not possible to implement high-quality growth of mono-cristalline samples on suitable surfaces, all the potenatial applications remain in the realm of "dreams" with no possibility to move to reality.

References

[1]   https://en.wikipedia.org/wiki/Flash_memory
[2]   https://www.nobelprize.org/prizes/physics/2007/fert/facts/
[3]   Prati, E. J Nanosci Nanotechnol. 11, 10, 2011, 8522–8526(5), American Scientific Publishers, https://doi.org/10.1166/jnn.2011.4957.
[4]   Thompson, S. E. et al., A 90-nm logic technology featuring strained-silicon. IEEE Trans. Electron Devices. 51, 2004, 1790–1797, https://10.1109/TED.2004.836648.
[5]   Sham, L., Allen, S. Jr, Kamgar, A., Tsui, D. Valley– Valley splitting in inversion layers on a high-index surface of silicon. Phys. Rev. Lett. 40, 472, 1978, https://doi.org/10.1103/PhysRev Lett.40.472.
[6]   Cole, T., Lakhani, A. A., Stiles, P. J. Phys. Rev. Lett. 38, 1977, 722. https://doi.org/10.1103/ PhysRevLett.38.722.
[7]   Ohkawa, J. F., Uemura, Y. Theory of valley splitting in an N-channel (100) inversion layer of Si I. Formulation by extended zone effective mass theory. J. Phys. Soc. Jpn. 43, 1977, 907–916. https://doi.org/10.1143/JPSJ.43.907.
[8]   Shkolnikov, Y., De Poortere, E., Tutuc, E., Shayegan, M. Valley splitting of AlAs two-dimensional electrons in a perpendicular magnetic field. Phys. Rev. Lett. 89, 2002, 226805. https://doi.org/10.1103/PhysRevLett.89.226805.
[9]   Isberg, J., Gabrysch, M., Hammersberg, J. et al., Generation, transport and detection of valley-polarized electrons in diamond. Nature Mater. 12. 2013, 760–764. https://doi.org/10.1038/ nmat3694.

[10] Koiller, B., Hu, X., Das Sarma, S. Exchange in silicon-based quantum computer architecture. Phys. Rev. Lett. 88, 2001, 027903. https://doi.org/10.1103/PhysRevLett.88.027903.

[11] Goswami, S., Slinker, K., Friesen, M. et al., Controllable valley splitting in silicon quantum devices. Nature Phys. 3. 2007, 41–45. https://doi.org/10.1038/nphys475.

[12] Yang, C., Rossi, A., Ruskov, R. et al., Spin-valley lifetimes in a silicon quantum dot with tunable valley splitting. Nat. Commun. 4, 2069, 2013, https://doi.org/10.1038/ncomms3069.

[13] Salfi, J., Mol, J., Rahman, R. et al., Spatially resolving valley quantum interference of a donor in silicon. Nature Mater. 13. 2014, 605–610. https://doi.org/10.1038/nmat3941.

[14] Schaibley, J. Chapter 10 × Valleytronics in 2D semiconductors, Editor(s): Bao, Q., Hoh, H. Y., Woodhead Publishing Series in Electronic and Optical Materials, 2D Materials for Photonic and Optoelectronic Applications. 2020, Woodhead Publishing, 281–302. ISBN 9780081026373 https://doi.org/10.1016/B978-0-08-102637-3.00010-3.

[15] Xiao, D., Yao, W., Niu, Q. Valley-Contrasting Physics in Graphene: Magnetic Moment and Topological Transport. Phys. Rev. Lett. 99, 2007, 236809. https://doi.org/10.1103/PhysRev Lett.99.236809.

[16] Xiao, D., Liu, G.-B., Feng, W., Xu, X., Yao, W. Coupled Spin and Valley Physics in Monolayers of MoS2 and Other Group-VI Dichalcogenides. Phys. Rev. Lett. 108, 2012, 196802. https://doi.org/10.1103/PhysRevLett.108.196802.

[17] Schaibley, J., Yu, H., Clark, G. et al., Valleytronics in 2D materials. Nat Rev Mater. 1. 2016, 16055. https://doi.org/10.1038/natrevmats.2016.55.

[18] Yao, W., Xiao, D., Niu, Q. Valley-dependent optoelectronics from inversion symmetry breaking. Phys. Rev. B. 77, 2008, 235406. https://doi.org/10.1103/PhysRevB.77.235406.

[19] Xiao, D., Yao, W., Niu, Q. Valley-contrasting physics in graphene: Magnetic moment and topological transport. Phys. Rev. Lett. 99, 2007, 236809. https://doi.org/10.1103/PhysRev Lett.99.236809.

[20] Xiao, D., Liu, G.-B., Feng, W., Xu, X., Yao, W. Coupled spin and valley physics in monolayers of MoS2 and other group-VI dichalcogenides. Phys. Rev. Lett. 108, 2012, 196802. https://doi. org/10.1103/PhysRevLett.108.196802.

[21] Xu, X., Yao, W., Xiao, D. et al., Spin and pseudospins in layered transition metal dichalcogenides. Nature Phys. 10. 2014, 343–350. https://doi.org/10.1038/nphys2942.

[22] Gong, Z., Liu, G. B., Yu, H. et al., Magnetoelectric effects and valley-controlled spin quantum gates in transition metal dichalcogenide bilayers. Nat. Commun. 4. 2013, 2053. https://doi. org/10.1038/ncomms3053.

[23] Xu, X., Yao, W., Xiao, D. et al., Spin and pseudospins in layered transition metal dichalcogenides. Nature Phys. 10. 2014, 343–350. https://doi.org/10.1038/nphys2942.

[24] Jones, A., Yu, H., Ross, J. et al., Spin–layer locking effects in optical orientation of exciton spin in bilayer WSe2. Nature Phys. 10. 2014, 130–134. https://doi.org/10.1038/nphys2848.

[25] Liu, G.-B., Xiao, D., Yao, Y., Xu, X., Yao, W. Electronic structures and theoretical modelling of two-dimensional group-VIB transition metal dichalcogenides. Chem. Soc. Rev. 44, 2015, 2643–2663. https://doi.org/10.1039/C4CS00301B.

[26] Consadori, F., Frindt, R. Crystal size effects on the exciton absorption spectrum of WSe2. Phys. Rev. B. 2, 1970, 4893. https://doi.org/10.1103/PhysRevB.2.4893.

[27] Frindt, R. Optical Absorption of a Few Unit-Cell Layers of MoS2. Phys. Rev. 140, 1965, A536. https://doi.org/10.1103/PhysRev.140.A536.

[28] Frindt, R., Yoffe, A. Physical properties of layer structures: optical properties and photoconductivity of thin crystals of molybdenum disulphide. Proc. R. Soc. Lond. Ser. A. Math. Phys. Sci. 273, 1963, 69–83. https://doi.org/10.1098/rspa.1963.0075.

[29] Mak, K. F., Lee, C., Hone, J., Shan, J., Heinz, T. F. Atomically thin MoS2: A new direct-gap semiconductor. Phys. Rev. Lett. 105, 2010, 136805. 10.1103/PhysRevLett.105.136805.

[30] Splendiani, A., Sun, L., Zhang, Y., Li, T., Kim, J., Chim, C. Y., Galli, G., Wang, F. Emerging photoluminescence in monolayer MoS2. Nano Lett. 10, 4, 2010Apr14, 1271–1275. 10.1021/nl903868w.

[31] Mak, K. F., Lee, C., Hone, J., Shan, J., Heinz, T. F. Atomically Thin MoS2: A New Direct-Gap Semiconductor. Phys. Rev. Lett. 105, 2010, 136805. https://doi.org/10.1103/PhysRevLett.105.136805.

[32] Knox, R. S. Theory of excitons, Solid state physics. Ed. by Seitz, Turnbul, Academic, NY Vol. 5, 1963, Liang, W Y 1970.

[33] Combescot, M., Shiau, S.-Y. Excitons and Cooper Pairs: Two Composite Bosons in Many-Body Physics. Oxford University Press, ISBN 9780198753735.

[34] Yao, W., Xiao, D., Niu, Q. Valley-dependent optoelectronics from inversion symmetry breaking. Phys. Rev. B. 77, 2008, 235406. https://doi.org/10.1103/PhysRevB.77.235406.

[35] Xiao, D., Liu, G.-B., Feng, W., Xu, X., Yao, W. Coupled spin and valley physics in monolayers of MoS2 and other group-VI dichalcogenides. Phys. Rev. Lett. 108, 2012, 196802. https://doi.org/10.1103/PhysRevLett.108.196802.

[36] Yao, W., Xiao, D., Niu, Q. Phys. Rev. B. 77, 2008, 235406. https://doi.org/10.1103/PhysRevB.77.235406.

[37] Giovannetti, G. et al., Substrate-induced band gap in graphene on hexagonal boron nitride: Ab initio density functional calculations. Phys. Rev. B. 76, 2007, 073103. https://doi.org/10.1103/PhysRevB.76.073103.

[38] Zhou, S., Gweon, G. H., Fedorov, A. et al., Substrate-induced bandgap opening in epitaxial graphene. Nature Mater. 6. 2007, 770–775. https://doi.org/10.1038/nmat2003.

[39] Mak, K., He, K., Shan, J. et al., Control of valley polarization in monolayer MoS2 by optical helicity. Nature Nanotech. 7. 2012, 494–498. https://doi.org/10.1038/nnano.2012.96.

[40] Xiao, D., Liu, G. B., Feng, W., Xu, X., Yao, W. Coupled spin and valley physics in monolayers of MoS2 and other group-VI dichalcogenides. Phys. Rev. Lett. 108, 2012, 196802. https://doi.org/10.1103/PhysRevLett.108.196802.

[41] Zhu, Z. Y., Cheng, Y. C., Schwingenschlogl, U. Giant spin-orbit-induced spin splitting in two-dimensional transition-metal dichalcogenide semiconductors. Phys. Rev. B. 84, 2011, 153402. https://doi.org/10.1103/PhysRevB.84.153402.

[42] Cheiwchanchamnangij, T., Lambrecht, W. R. L. Quasiparticle band structure calculation of monolayer, bilayer, and bulk MoS2. Phys. Rev. B. 85, 2012, 205302. https://doi.org/10.1103/PhysRevB.85.205302.

[43] Mak, K., He, K., Shan, J. et al., Control of valley polarization in monolayer MoS2 by optical helicity. Nature Nanotech. 7. 2012, 494–498. https://doi.org/10.1038/nnano.2012.96.

[44] Jones, A., Yu, H., Ghimire, N. *et al.*, Optical generation of excitonic valley coherence in monolayer WSe$_2$. Nature Nanotech. 8. 2013, 634–638. https://doi.org/10.1038/nnano.2013.151.

[45] Alexey Chernikov, T. C., Berkelbach, H. M., Hill, A. R., Yilei, L., Aslan, O. B., Reichman, D. R., Hybertsen, M. S., Tony, F. H. Phys. Rev. Lett. 113, 2014, 076802.

[46] Mak, K., He, K., Shan, J. et al., Control of valley polarization in monolayer MoS2 by optical helicity. Nature Nanotech. 7. 2012, 494–498. https://doi.org/10.1038/nnano.2012.96.

[47] Cao, T., Wang, G., Han, W. et al., Valley-selective circular dichroism of monolayer molybdenum disulphide. Nat. Commun. 3. 2012, 887. https://doi.org/10.1038/ncomms1882.

[48] Zeng, H., Dai, J., Yao, W. et al., Valley polarization in MoS2 monolayers by optical pumping. Nature Nanotech. 7. 2012, 490–493. https://doi.org/10.1038/nnano.2012.95.

[49] Vitale, S. A., Daniel Nezich, J. O., Varghese, P. K., Gedik, N., Jarillo-Herrero, P., Xiao, D., Rothschild, M. Valleytronics: Opportunities, Challenges, and Paths Forward. Small. 14, 2018, 1801483. http://doi.org/10.1002/smll.201801483.

[50]   https://qt.eu/discover-quantum/underlying-principles/quantum-key-distribution-qkd/

[51]   Zhao, S., Xiaoxi, L., Dong, B., Wang, H., Wang, H., Zhang, Y., Han, Z., Zhang, H. Valley
       manipulation in monolayer transition metal dichalcogenides and their hybrid systems: Status
       and challenges. Rep. Prog. Phys. 84, 2, 2021, O26401. https://iopscience.iop.org/article/10.
       1088/1361-6633/abdb98.

[52]   https://research.a-star.edu.sg/articles/features/on-the-cusp-of-valleytronics/

[53]   https://insidehpc.com/2021/09/indian-institute-of-tech-max-born-institut-team-say-gra
       phene-valleytrionics-could-enable-room-temperature-quantum/

4 2D black phosphorus: difficult to handle but so interesting for opto-electronics

4.1 Introduction

Phosphorus (P) is one of the most abundant elements that can be found on the Earth considering that around 0.1% of the Earth's crust is composed of it [1, 2]. P, as in case of carbon, can be under several allotropic forms such as white phosphorus, red phosphorus, black phosphorus (BP), violet phosphorus and A7 phase. In Fig. 4.1, the various allotropic forms of phosphorus are presented [3]. As highlighted by Carvalho et al. in their very interesting review [4], the first appearance of phosphorus dated around 3,000 years ago in ancient China and was spotted through the spontaneous combustion of P_2H_2. Indeed, elemental phosphorus was isolated for the first time by Brand in 1669. This last, an alchemist researching the philosophical stone, focused his work on water trying to combine it with different other materials, in all potential combinations. In 1669, he tried heating residues from boiled down urine [5] on his furnace until the retort was red hot. Suddenly, glowing fumes filled it and liquid dripped out, bursting into flames. After recovering the liquid in a jar and covered it, this last gets solidified and continued to emit a pale-green glow. Indeed, what he was able to collect was a material that was called phosphorus, using the Greek word meaning "light-bearing" or "light-bearer". Therefore, we can affirm that the discovery of phosphorus relates to a serendipity process. Even if the discovery of these materials was not really a "glamour" event, the expectances for this material are extremely promising.

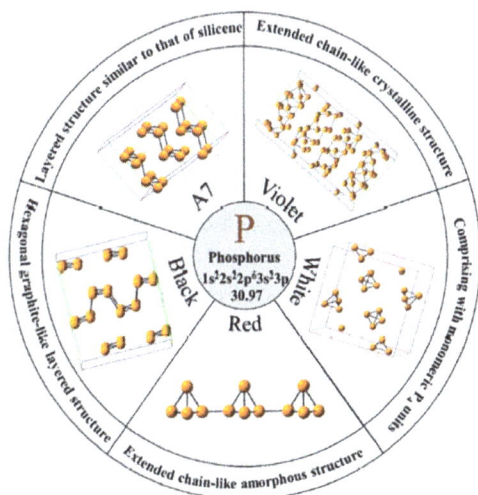

Fig. 4.1: Allotropic forms of phosphorus (reproduced with permission [2], Copyright 2017, American Chemical Society).

https://doi.org/10.1515/9783110656336-005

4.2 Different allotropic forms and structures of phosphorus

Our attention will be focused on BP. BP cannot be called a "new material" considering that it has been synthesized for the first time in 1914 in its bulk form by Bridgman [6]. Bridgman successfully transformed white phosphorus in BP, submitting the first one at a pressure of 1.2 GPa and at a temperature of around 200 °C. The main difference between these two different allotropic forms is that the BP was stable at ambient conditions, in opposite to white and red phosphorous. In total, at least five crystalline polymorphs of P and several amorphous forms have been synthesized and others have been theoretically predicted [7–9]. Since the late 1960s, the booming interest in superconductivity headed some scientific teams to focus their attention on the rich phase diagram of phosphorus. Actually, it was observed that P could potentially own a superconducting phase at a critical temperature, T_c, of 4–10 K, when it is in its high-pressure cubic and rhombohedral forms [10–13]. This immediately pushed for some years to study deeply high-pressure routes for BP growth. However, important bottlenecks were quickly identified which led to slow down strongly the research in the field: the difficulty in synthesizing in a reproducible way the material and its bad quality. If we do not consider some anecdotic works, no particular interest was shown by the scientific community and the research in the field for around 100 years. Indeed, Bridgman was awarded the Nobel Prize in 1946 for "the invention of an apparatus to produce extremely high pressures, as well as the discoveries he made therewith in the field of high-pressure physics" [14] which did not deal specifically with the BP properties of the material by itself. To be precise, as outlined by Ling et al. [15] in an extremely interesting paper of 2014 on the reborn scientific interest on BP, in 100 years only around 100 papers were written on BP, a very low number compared to the number of publications in the last years (more than 1,000). Some scientific groups, mainly in Japan in the 1970s and 1980s, continued, "under the radar", to study the structure [16, 17], the transport[18], the optical [19–22], electrical [23, 24] properties [25, 26] and potential superconductivity [27]. These early studies did not receive much attention from the semiconductor research community at that time and in the following years, because we were in the booming years of development of silicon-based devices, the so-called gold age of Moore's law. Only recently, with deeper studies on the fundamental properties and because of the hype of 2D materials, BP was recently rediscovered. The immediate consequence was a huge interest from scientists from different backgrounds such as condensed matter physicists, chemists, semiconductor device engineers and material scientists. Indeed, this is a similar story to what happened for graphene. In fact, the Nobel Prize in 2010 given to Geim and Novoselov was not about the discovery of graphene, already theoretically predicted in the 1950s and observed in the 1960s [28], but more precisely "for ground-breaking experiments regarding the two-dimensional material graphene" [29]. As already seen for graphite and transition metal dichalcogenides (TMDs), BP owns a layered structure with unique puckered single-layer geometry. Analysing more precisely the BP structure, we can define BP as a single-elemental-layered crystalline material consisting of only phosphorus atoms. The main difference with other 2D-layered materials of group IV, such as graphene, silicene or germanene, is that each phosphorus atom has five outer shell electrons. BP presents three crystalline structures, as already presented and highlighted by Morita et al. [30] and Scelta et al. [31]: orthorhombic, simple cubic and rhombohedral as a function of the phase diagram (see Fig. 4.2).

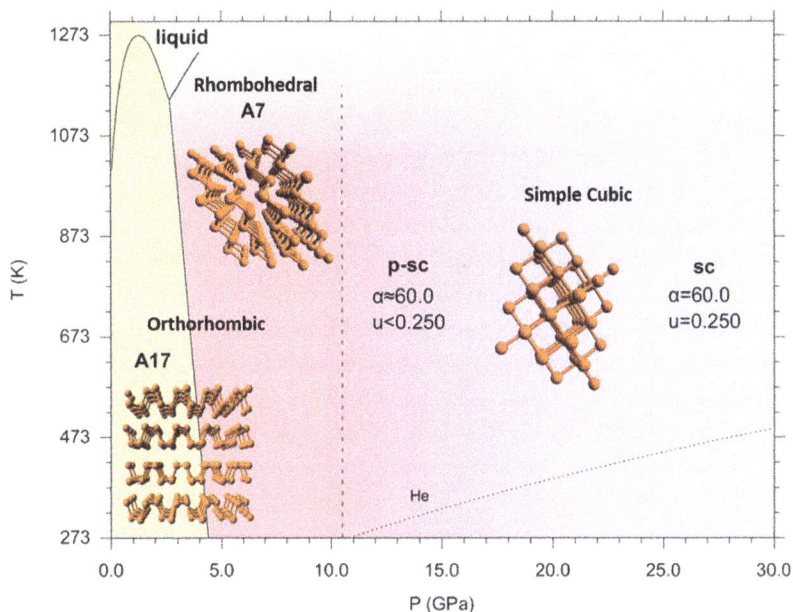

Fig. 4.2: Crystalline structures of BP as a function of the temperature and pressure (reprinted with permission from John Wiley and Sons Inc, *Angewandte Chemie International Edition* [31], Copyright John Wiley and Sons Inc. (2017)).

Semiconducting puckered orthorhombic BP belongs to the D_{18}^{2h} point group. This group shows a reduced symmetry if compared to group IV components (such as graphene) owing to the D_4^{6h} point group symmetry [32]. The single-layer BP is composed of two atomic layers and two kinds of P–P bonds, respectively, the shorter bond length of 0.2224 nm, linking the nearest P atoms in the same plane, and the longer bond length of 0.2244 nm linking P atoms between the top and bottom of a single layer. The top view of BP along the z-direction shows a non-perfect hexagonal structure with different bond angles, highlighted by two main works of Asahina et al. in 1982 and Takao et al. in 1981: 96.3° and 102.1° [33, 34].

The renewed interest for BP has been emerging since 2014 after the exploding interest of the scientific community for 2D materials with morphological characteristics similar to graphene, such as hexagonal boron nitride and TMDs (e.g. MoS_2).

For this reason, the main studies of scientific teams have been focusing mainly in the perspectives of using single-layer (also called phosphorene), few layered materials and thin films and to disclose their peculiar characteristics.

4.3 Why BP is so interesting? Tailoring the band gap through thickness

Coming back to the BP structure, we can observe that P atoms are strongly bonded in plane forming layers (see Fig. 4.3), while different layers staying compacted, thanks to interlayer weak van der Waals (vdW) forces. A single layer of bulk BP is usually obtained through mechanical exfoliation as exactly in the case of graphene from graphite. The bulk BP has an orthorhombic structure as shown in Fig. 4.3. Therefore, each phosphorus atom is bound to three neighbouring atoms through sp^3-hybridized orbitals, leading to a puckered honeycomb structure. Consequently, an important feature of BP is that each atom has a lone electron pair, and the remaining lone pairs make phosphorus reactive to air. This is an important property of BP that does not allow handling it in a common way. As outlined by Wu et al. [36], bulk BP is a p-type semiconductor owing a direct band gap, showing good electrical conductivity ($\approx 10^2$ S/m), reasonable density (2.69 g/cm^3) and an intrinsic energy gap of around 0.34 eV [37]. The semiconductor state has interesting electrical properties with an electron and hole mobility, respectively, of 220 and 350 cm^2/V s [38]. Indeed BP, as shown in Fig. 4.2, because of Bridge's experiments, shows three crystalline phases, more specifically orthorhombic, rhombohedral and simple cubic phases. These phases are strictly associated with different pressure values that change dramatically the

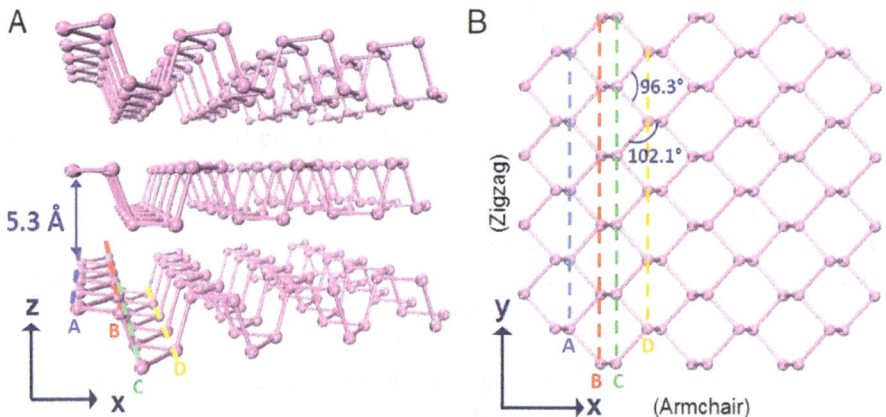

Fig. 4.3: Crystal structure and band structure of BP. (A) Side view of the black P crystal lattice with interlayer spacing of 0.53 nm. (B) Top view of the lattice of single-layer black P with bound angles [35] (reprinted with permission from [35], Copyright (2015) National Academy of Sciences.

electronic characteristics of BP. We observe that under high pressure (5.5 GPa), semi-conducting orthorhombic BP, stable under ambient conditions, can be transformed into a semi-metallic rhombohedral structure [39]. As already highlighted, considering that BP can display two types of bonds, namely, covalent intralayer bonding and weak interlayer vdW bonding [40], increasing pressures, the distances among the stacked layers logically decrease faster than the intralayer atomic ones [41]. Employing a much higher pressure (10 GPa), Ahuja et al. [42] observed that the semi-metallic rhombohedral structure finally changed itself into metallic cubic phase because of an internal distortion. Summarizing, as previously stated, the difficulties in synthesizing BP are one of the most important reasons that discouraged the research efforts on this material for around 100 years. This until a BP single layer (a.k.a. "phosphorene") was finally isolated successfully by using the scotch-tape technique used to isolate graphene [43]. This gave a boost to the research in the field. In the impressing work of Liu et al. [44] in 2014, scientists not only succeeded in isolating one single layer of phosphorene and demonstrating, after simulations and experiments, that the direct band gap of the material was quite larger than that already found for the BP as shown in Fig. 4.4.

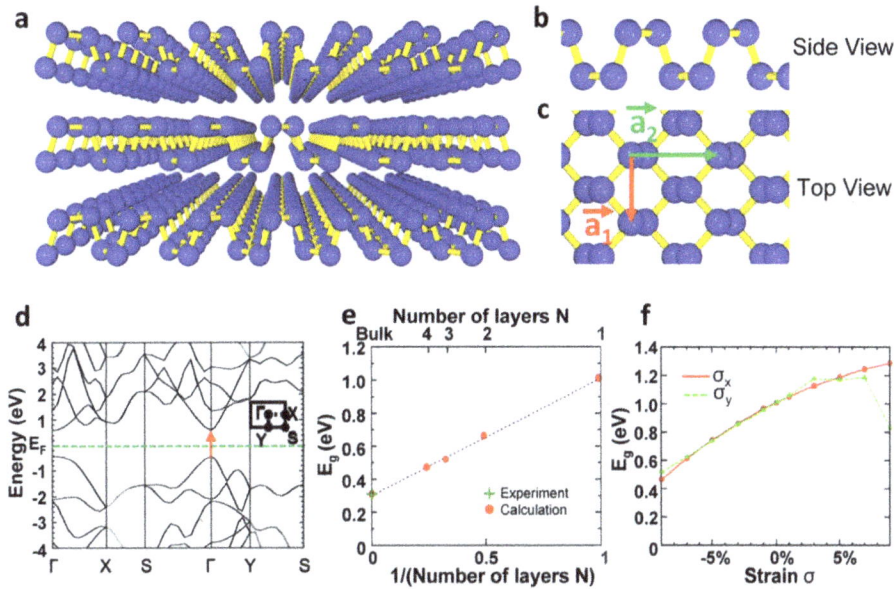

Fig. 4.4: Crystal structure and band structure of few layer phosphorene. (a) Perspective side view of few layer phosphorene. (b) and (c) Side and top views of few layer phosphorene. (d) Band structure of a phosphorene monolayer. (e) and (f) Energy gap of the samples as a function of layers and uniaxial strain (e) (reprinted with permission from [44], Copyright (2014) American Chemical Society).

a

b

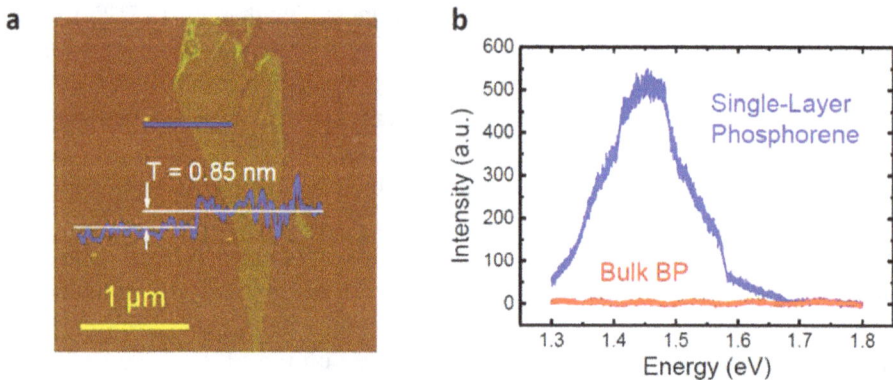

Fig. 4.5: Characterizations of single-layer and few layer BP. (a) Atomic force microscopy image of a single-layer phosphorene crystal with the measured thickness of ≈0.85 nm. (b) PL spectra for single-layer phosphorene and bulk BP (reprinted with permission from [44], Copyright (2014) American Chemical Society).

This was demonstrated, thanks to photoluminescence tests that pointed out a peak in the visible wavelength from single-layer phosphorene confirming indirectly the widening of the band gap as predicted by theory (see Fig. 4.5). As put in the evidence in the ab initio simulations performed by of Li and co-workers, a very interesting feature of this material is that the band gap value is strictly dependent on the number of layers in the sample. As shown in Fig. 4.4(d), the calculated band structure outlines that a free-standing phosphorene single layer is intrinsically a direct band gap semiconductor with energy gap value of 1.0 eV at Γ, which is clearly larger than the band gap calculated for bulk BP. A peculiar characteristic of this material is that its band gap value, E_g, changes in an inversely proportional way to the number of layers stacked and varies between 1.0 eV for a single layer and 0.3 eV in the bulk. This feature outlines the possibility to modulate the electronic properties of the material achieving ad hoc opto-electronic properties. On the other side, the gap is also dependent, as shown in Fig. 4.4(f), by the in-plane strain. For example, because of a moderate in-plane uniaxial strain (compression) of ≈5%, Li et al. observed that the direct band gap of the material muted in indirectly. This kind of strain can be originated in case of growth on specific substrates with a mismatch, which emphasizes the importance of the grown substrate in case of implementation in real devices. Experiments performed in the framework of the same study, and specifically photoluminescence measurements, demonstrated that the band gap of phosphorene was larger than the bulk one. Indeed, if we compare the results obtained by Li et al., the peak at 1.45 eV found by the researcher was identified as lower energetic excitonic bound. This gives us an idea of the lower estimation of the phosphorene band gap that will be large than that. These results were also confirmed by a work of Tran et al. in 2014 [45]. The difference between the theoretical values and the

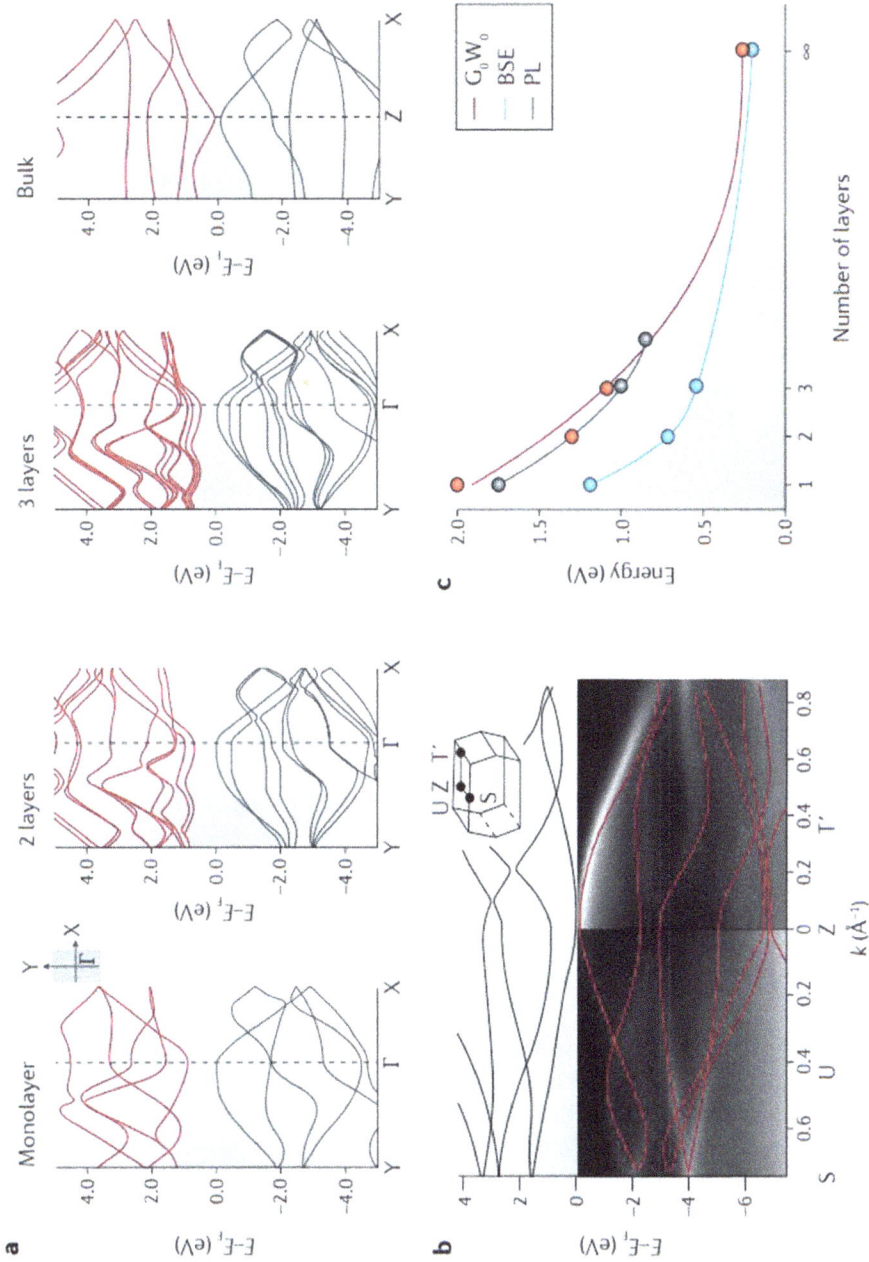

Fig. 4.6: Band structure of phosphorene. (a) Band structures obtained by DFT for different layers. (b) Angle-resolved photoemission spectroscopic spectra compared with bands obtained with DFT calculations. The inset shows the first 3D Brillouin zone of black phosphorus. (c) Band gap as a function of the number of layers, theory and experimental results (reprinted with permission from Springer Nature Customer Service Centre GmbH, Springer Nature [47], Copyright Springer (2016)).

experiments could be due to potential impurities in the materials or to the fact that theoretical predictions had been conceived as the sample was in vacuum.

As shown in Fig. 4.6(a), the shapes of the band gaps of BP in case of phosphorene or multi-layered structure show the same profile because they have the same translational structural symmetry and bond interactions [34, 46, 47].

Indeed, we can observe that the band gap remains direct at the Γ point of the Brillouin zone. As pointed out by Carvalho et al. [4], when we observe the band structure of the monolayer, bilayer and trilayer, we notice a redshift of the band gap as a function of the increasing number of layers and a contemporaneous splitting of the bands. However, in an unexpected way, and not like in case, for example, of some 2D TMD materials such as MoS_2, the band dispersion is not strongly affected and remains at the same Γ point of the Brillouin zone for the different samples composed of different layers. This takes place because the top of the valence band is quite flat and so a very shallow change of the energy value of the maximum has no consequence and can be neglected considering that it cannot be measured at ambient conditions. As previously outlined, the presence of a direct band gap for any number of layers is a peculiar characteristic if compared to MoS_2 or WS_2, which display an indirect-to-direct band gap transition when we move from bulk to monolayer material. If we consider these peculiar characteristics from the point of view of potential implementation of material in the field of opto-electronics, technologically it will surely be easier to integrate thicker stable samples than single-layer phosphorene, considering that all these materials own a direct band gap. An opportunity is also emerging about the utilization of BP-based materials that can span the spectral region, from visible to the middle infrared not covered by commonly used TMD semiconductors. We will discuss that in a more precise way in a following section devoted to potential applications.

4.4 Passivation

The main drawback in case of implementation of BP is the extremely rapid and quite spectacular degradation of the material when in contact with air [48]. In fact, the exfoliated layers of phosphorene are subject to the oxidation process happening immediately upon exposure [49]. Some parameters are very important such as the presence of humidity and light, which lead to the formation of PO_x [48]. For example, some theoretical works [50] demonstrated that BP has a strong dipolar moment out of plane, which leads to a strong hydrophilic character of the surface. Therefore, we observe the formation of droplet on the surface of single-layer BP that can be simply related to adsorbed or condensed environmental moisture on the surface. This phenomenon leads to the charge transfer from phosphorus atoms to oxygen ones and consequent synthesis of phosphoric acid, deteriorating the sample surface [51]. The direct consequence is the strong deterioration of the intrinsic physical properties of the material. Therefore, the capacity of stabilizing phosphorene is fundamental to figure its potential industrial implementation. It has been observed that the quality of phosphorene layer reduces against the size, and the degradation severity depends on the layer thickness because of the quantum-confinement effects [52]. The main reason is that reducing the layer thickness the gap is widened, moving the extremities of the band-gap to larger values of energy and so leading to the alignment of the valence band maximum and conduction band minimum with that of oxygen. [53]. Thanks to that phosphorene which shows a lower reactivity compared to other elemental 2D materials like silicene and can be handled at ambient conditions for a certain duration of time, even if quite limited. Enhancing the stability of BP in ambient conditions will be a fundamental key to allow these materials to be integrated in real devices and to impact real applications. In the following paragraphs, we will summarize some of the main technologies and strategies to improve BP stability in ambient conditions. It is not our objective to give an exhaustive analysis, which is already provided by several review on the matter; however, we want only to give a global view and to highlight some interesting contributions with higher potential for implementation. Different strategies have been developed to tackle this problem in case of 2D BP such as encapsulation, functionalization, liquid phase surface passivation and doping. In this paragraph, we will try to explain in a simple way which are the main challenges to apply in a suitable way these techniques.

4.4.1 Encapsulation

In case of 2D BP, the encapsulation method has been largely exploited using different kinds of layers. Atomic layer deposition (ALD) is a highly scalable and manufacturable technique to prepare conformal and pinhole-free high-k dielectric thin films for gate insulator and diffusion barrier applications in conventional Si-based COMS devices, which involves the reaction of metal precursors and an oxidation agent, such as water or ozone, that reacts with the surface separately. Thanks to ALD, high-k dielectrics such as aluminium oxide (Al_2O_3), zirconium oxide (ZrO_2) or hafnium dioxide (HfO_2) can be deposited on 2D materials to perform effective encapsulation. One of the most effective techniques employed for 2D BP is ALD encapsulation of AlO_x layer. This method has been extensively explored to protect BP from air degradation. We can mention the very interesting work of Wood et al. in 2014 [54]. In this contribution, scientists observed that the BP layer degradation of BP field-effect transistors (FET) was at the origin of the large increase in threshold voltage after 6 h in ambient conditions. As a consequence, the FET on/off ratio followed by a ~10^3 decrease in FET current on/off ratio and mobility after 48 h (see Fig.4.7).

Thanks to encapsulation using ALD AlO_x, scientists could dramatically reduce ambient deterioration, allowing encapsulated BP FETs to continue to show high on/off ratios of ~10^3 and quite good mobilities, even if not excellent compared to

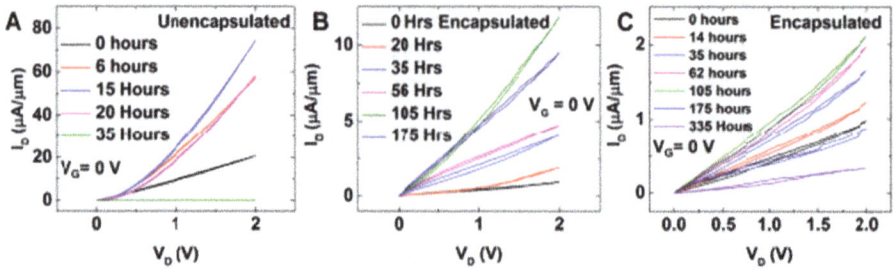

Fig. 4.7: *I–V* curves of BP FET unencapsulated and encapsulated that puts in evidence the strong effect of encapsulation on the lifetime of devices (reprinted (adapted) with permission from [54], Copyright (2014) American Chemical Society).

other 2D materials, of ~100 cm^2/V s for over 2 weeks in ambient conditions. Illarionov et al. reported in 2017 that BP FETs with conformal 25 nm Al_2O_3 encapsulation could extend this time and showed a highly stable performance up to ~17 months [55]. The efficient encapsulation using AlO_x was also confirmed successively by another work by Galceran and co-workers in 2017 [56]. In this work, several characterization techniques such as optical microscopy, Raman spectroscopy and atomic force microscopy studies were employed to verify the efficiency of a 1 nm thick Al_2O_3 layer passivating exfoliated few layered BP (<5 layers) on Si/SiO_2 substrates. In this case, we can talk about an ultrathin and transparent passivation layer acting also as a tunnel barrier allowing BP-based devices processing without passivation layer removal. Two years later, scientists of the same team [57] developed a new process to create passivation barrier for BP acting as a spin transport channel and, at the same time, as tunnel barrier necessary to achieve spin injection. This time, they developed a large surface passivation for in situ ALD deposition of a continuous barrier of MgO that allowed protecting BP flakes for at least 1 month at ambient condition (see Fig. 4.8), which is a major breakthrough.

4.4.2 Functionalization

Another approach to avoid the deterioration of BP and to allow the implementation of this material in a real device for electronic or opto-electronic applications [58–63] is to use different types of organic or inorganic nanomaterials through covalent or non-covalent bonding. Indeed, thanks to this strategy of encapsulation through functionalization, two tasks can be filled only using one single process. Firstly, the passivation and so the protection of BP flakes against environmental agents but also, in a second stance, the modulation of charge-transfer characteristics between different 2D materials, for example, in vdW structures [64] or also in transistors. In a work by Gao et al. in 2012 [65], scientists studied the behaviour of single-layer

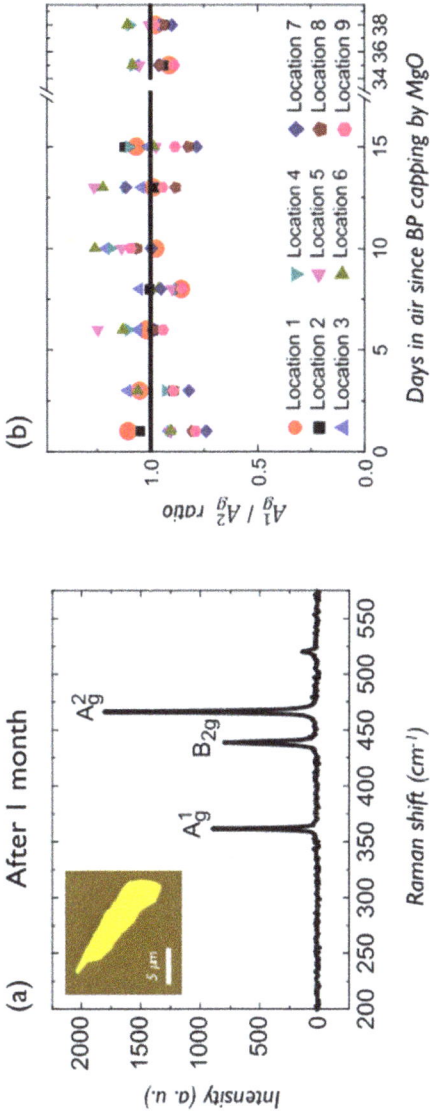

Fig. 4.8: Raman characterization of MgO passivated flakes after extended period in air. (a) BP Raman signatures after 1 month in air. (b) Ratio of intensity of peaks as a function of location on the sample surface and upon exposure (reprinted with permission from [57], Copyright (2019) AIP Publishing).

phosphorene in vacuum integrated in vdW heterostructures based on MoSe$_2$. In a quite unexpected way, researchers discovered that when phosphorene was overlying the vdW heterostructures and undergoing a suitable vertical electric field, the phosphorene was preserved by deterioration at ambient conditions, reaching up to 1×10^7 s of lifetime. This result seems to be related to the fact that the electric field permitted the manipulation of the relative energetic position of the band in oxygen and phosphorene, increasing the energy barrier for oxidation and so inhibiting the interaction with environmental oxygen and moisture. In the same contribution, scientists performed the calculation on the potential increase of lifetime and obtained that it could be amplified by a factor of 10^5 compared to bare BP. In case of covalent functionalization of the surface, in this case, we need to have some chemical reaction between the chosen functional group and the surface of the 2D material that has to be passivated. The main strategy is to inhibit chemically the reactivity of lone pair of electrons in BP, which is extremely reactive at ambient condition in the presence of oxygen molecules [66]. One of the most successful methods found in the literature is the functionalization using aryl diazonium (AD) molecules developed by Ryder and co-workers in 2016 [67] (see Fig. 4.9). These materials are able to extend the lifetime of phosphorene-based devices up to 3 weeks upon ambient air exposure As previously mentioned, this approach is very effective because AD alters the electronic properties of phosphorene achieving a chemical bond with phosphorene and pairing the lone phosphorene electrons that lose in this way their reactivity. Another advantage is that AD dopes the BP and increases the charge mobility in this last. This could be extremely useful in case of utilization of BP as a channel in FET also because the on/off ratio is enhanced.

Another approach exploiting non-covalent functionalization using perylene diimides (PDI) has been recently developed by Lloret et al. in 2020 [68]. Indeed in this case and inversely of what happened with AD that tended to be chemically bound to BP, the interaction of PDI with BP layers can be assimilated to an interlayer vdW interaction as shown in Fig. 4.10. The main researchers' hypothesis was based on the observation that enhancing the aromatic character of the periphery branches of PDI, it was easier to achieve a more effective molecular packing, increasing adsorption energies on BP surface and so leading to a more stable vdW interaction, and so a more effective protection of the BP layer. A similar result, using the same approach, has been obtained using perylenetetracarboxylic dianhydride on BP surface by Guo and co-workers in 2019 [69].

Lei and co-workers in 2018 [70] through DFT simulation studied the adsorption of different atoms such as Ca, Sr, Ba, Cs, La and Cl on the surface of BP flakes in order to reduce the energy of the bottom of the conduction band at a value lower than the oxygen redox potential. This technique, as the previous one, has an objective to inhibit the oxidation and extending the surface integrity lifetime. Finally, a similar approach that implements surface modification has been suggested by Guo et al. in

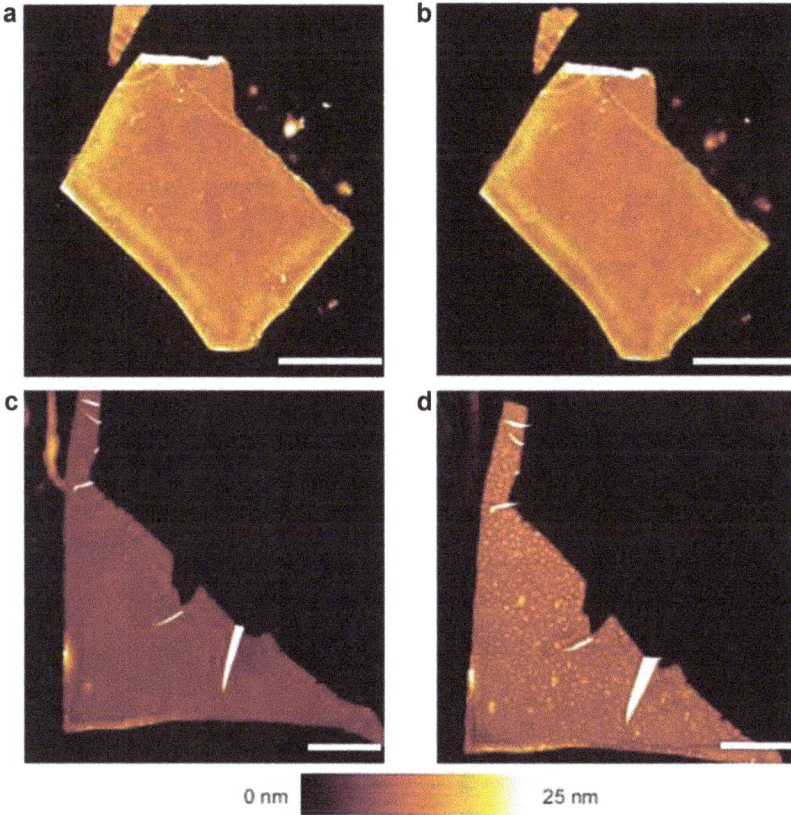

Fig. 4.9: Effect of functionalization on BP samples (AFM pictures) before and after ambient exposure. (a) Sample 1: BP flake immediately after functionalization. (b) The same flake (a)after 10 days of ambient exposure. (c) Sample 2: BP flake at the beginning of its exposure to air (d) after 10 days of ambient exposure showing degradation (reprinted with permission from Springer Nature Customer Service Centre GmbH, Springer Nature [67], Copyright Springer (2016)).

2017 [71]. In this case, researchers implemented the adsorption of Ag^+ metal ions on the surface of BP flakes. As for the other passivation technologies, also this strategy targets the lone pairs of electrons in P that tend to react with oxygen. Thanks to that scientists were able to obtain an extremely stable single-layer phosphorene as shown in Fig. 4.11. Another advantage, as already outlined by Ryder and co-workers [67], is the adsorption of elements that lead to p-dope the channel of potential BP FET and so to increase its mobility.

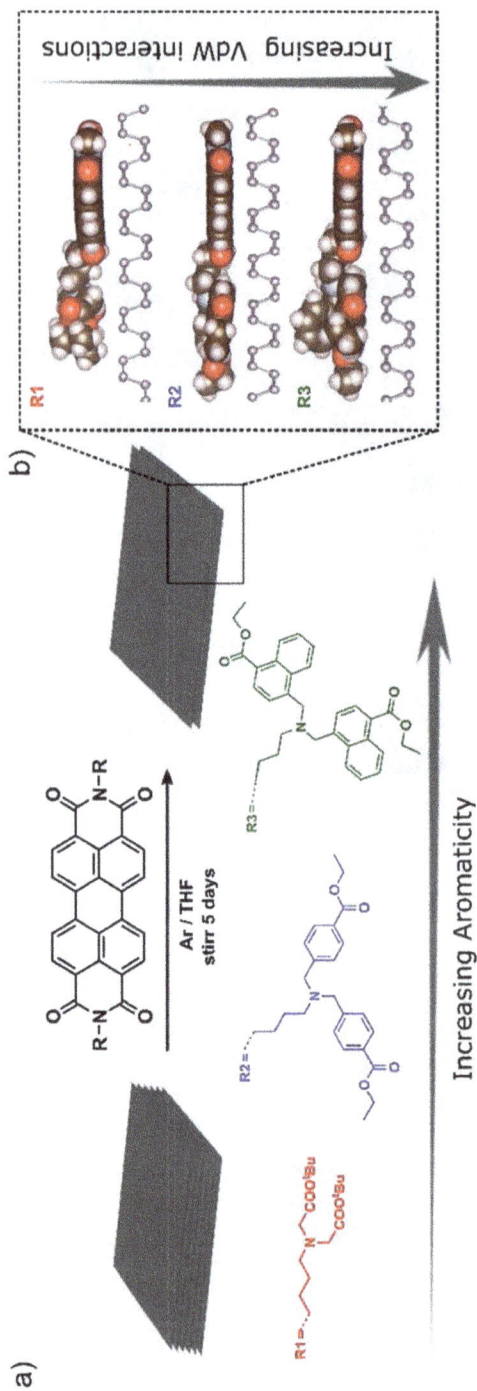

Fig. 4.10: (a) PDI functionalization of BP flakes with three different periphery branches. (b) DFT-optimized structures (reprinted with permission from John Wiley and Sons Inc, *Adv Mat. Interface* [68], Copyright John Wiley and Sons Inc. (2020)).

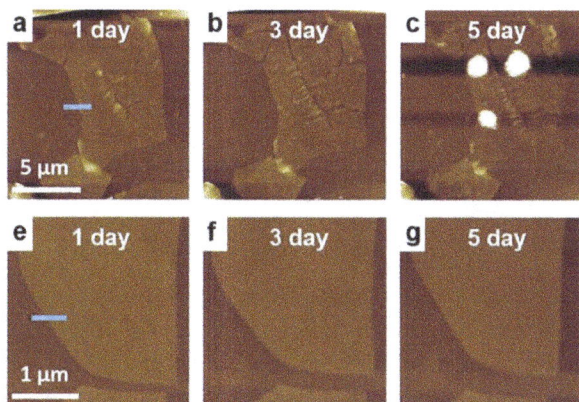

Fig. 4.11: Effect of functionalization using Ag$^+$ ions on BP. (a–d) AFM images of a bare BP sheet exposed to air for 1 day (a), 3 days (b) and 5 days (c). (e–h) AFM images (e–g) of a BPAg(+) sheet exposed to air for 1 day (e), 3 days (f) and 5 days (g) (reprinted with permission from John Wiley and Sons Inc, *Adv Mat.* [71], Copyright John Wiley and Sons Inc. (2017)).

4.4.3 Liquid-phase surface passivation

This is one of the common approaches that has been implemented at the very beginning for the passivation of BP flakes. This is quite logic considering that, as in case of graphene, the different teams wanted to develop a strategy to exfoliate single layer of BP. Two main methods have been often used: using ionic liquid (ILs) [72] or conductive polymers [73]. We can mention, as example, the work of Zhao and co-workers in 2015 [74]. Suspensions of BP flakes were performed using 1-hydroxyethyl-3-methylimidazolium trifluoromethanesulfonate showing a stability of around 1 month upon exposure to ambient air. However, in this case, even if scientists deal with dense suspensions with concentrations reaching around 1 g/L, we are talking about materials that remained in suspension and that were not deposed, for example, by spin-coating or integrated in a final device. Maybe the approach developed by Zhang et al. [75] in 2017 is more interesting, considering that during the developing of an effective process for exfoliation using ILs and specifically fluorination of the surface of the BP flakes in suspensions, they discovered that the fluorination enhanced stability at ambient condition after deposition of suspensions. The effect was double, as in the previous case the inhibition of the reactivity of the surface and, at the same time, the surface was made hydrophobic, thanks to that reducing strongly the effect of environmental moisture. In case of utilization of polymer IL compound, it has been proven that it might be another effective approach in surface passivation for few layered BP surface to avoid surface deterioration reaching up to 100 days. In this case, we are not discussing about the simple functionalization of materials but indeed about BP integrated in the matrix composed of polymer, so not exactly the same finality targeted with encapsulation or pure surface functionalization using, for example, adatoms or ions, where at the end we have a BP layer and not a composite. The final objective of this kind of approach is to create a final compound that can enable the improvement of performances, for example, for opto-electronics or for organic FETs (OFET). A good example is provided by the work of Pessaglia and co-workers in 2018 [76]. They reported the development of a process based on the formation of composite using polymer-based materials where again the stability of BP flakes upon exposure was improved, even in this case we cannot say that we deal with the direct exposure considering that we have a composite. The different steps to achieve the final composite were: (i) the mixing of liquid-phase exfoliated BP nanoflakes (using dimethyl sulfoxide) with poly(methyl methacrylate) solution; (ii) the exfoliation of BP through a polymeric solution; (iii) finally, the in situ radical polymerization, after performing exfoliating of BP, in MMA. Thanks to this process, the chemical structure of the BP nanoflakes was not deteriorated by ambient air and under UV light exposure. To end this section, we would like to mention the work of Brent et al. [77] in 2016. The main result of this contribution concerned the capacity of exfoliating BP multilayered samples using water-based suspensions with 1% in terms of weight of Triton X-100 (TX-100, $C_{14}H_{22}O(C_2H_4O)_n$, where $n = 9$–10). The function of this last compound was to create a sort of protective films for BP surface able to avoid, or at less to slow, the degradation in water as outlined in Fig. 4.12. This approach could be interesting in optics of using BP in applications dealing with biology or biochemistry.

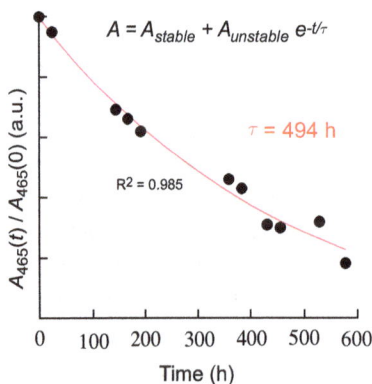

Fig. 4.12: UV–vis absorbance spectroscopy time study of the stability of few layered BP nanoflakes in 1% w/v aqueous Triton X-100. The change in absorbance at 465 nm is a parameter that can highlight the amount of black phosphorus remaining in solution (reprinted with permission from *RSC Advances* [77], Copyright Royal Society of Chemistry (2017)).

4.4.4 Doping

In some previous contribution, some teams developed approaches that firstly allowed the functionalization of the surface of BP and secondly promoted the doping of the materials, in this case, mainly p-type. Indeed, it has been demonstrated by some scientific contributions that doping can strongly prevent the deterioration of BP surface. We can quote the work of Wang et al. in 2019 [78], where they intentionally n-doped BP flakes through a simple thermal treatment. Wang and co-workers performed doping using Al adatoms that allowed the stabilization of 2D BP layers at ambient conditions and also improving the transfer characteristics of a BP FET built using the doped materials as a channel. As shown in Fig. 4.13, the layer remained unaltered after 10 days of exposure.

Yang and co-workers [79] doped BP flakes using tellurium increasing in a significant way the stability of the layer and also preserving in part the performances of BP FETs fabricated using doped layers. They were also able to increase the mobility of charges up to 1,850 cm^2/V s at ambient conditions. After 3 weeks, they observed a quite strong degradation of the performances and a final mobility of around 200 cm^2/V s, which is around 10% of the initial value. However, compared to non-protected samples, these last showed an immediate reduction to reach only 2% of the initial value. This approach made echoes to the previous example presented, considering that the Te atoms were used to pair up the lone dangling bond of P and so to reduce the Conductive band minimum (CBM) under the redox potential of O^2/O^{2-}. Thanks to that the generation of O^2 through light is dramatically reduced.

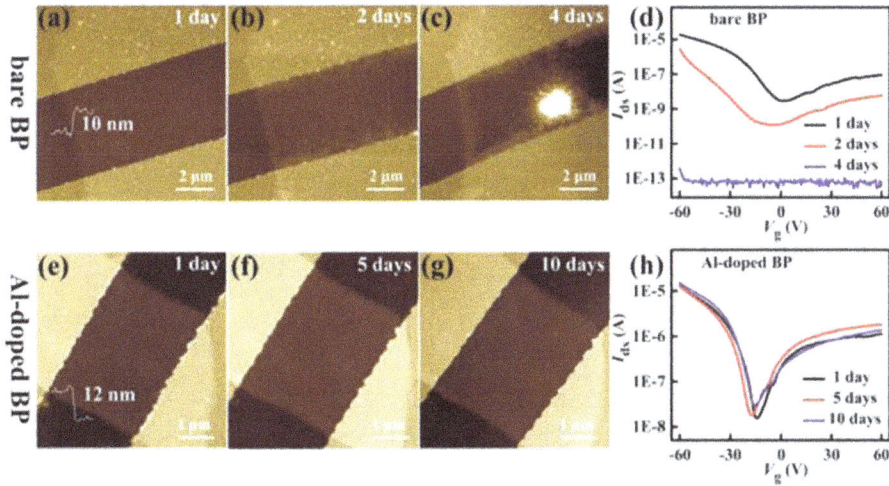

Fig. 4.13: (a)–c) Exposure of time-dependent AFM images and (d) transfer characteristics of the bare BP flake in air; (e)–(h) exposure of time-dependent AFM images and (h) transfer characteristics of the Al-doped BP flake in air (reprinted with permission from *Nanotechnology* [78], Copyright IOP (2019)).

4.5 Some examples of potential applications

4.5.1 Few layer BP in field-effect transistors

4.5.1.1 Introduction

Considering that few layered BP shows a gap, the first approach that many scientific teams followed, after 2014, has been to implement it on transistors. Our personal opinion on this specific topic remains the same. It is not useful to reproduce exactly the same devices, and particularly transistors, using different materials and this is mainly in an optics of publishing papers with a very limited benefit especially from the point of view of science and applications. It is clear that BP shows some interesting characteristics at the same time, which show the presence of a gap and the high mobility of charges. However, as in case of graphene, a real added value of 2D materials, is the possibility of disclosure new properties that will allow giving birth to a new generation of devices. Moreover, the existing transistor technology based on CMOS provides extremely miniaturized devices whose dimensions will be difficult to reach using these new 2D materials. For these reasons, even if the results were very promising, the industrial implementation of BP flakes in industrial FET

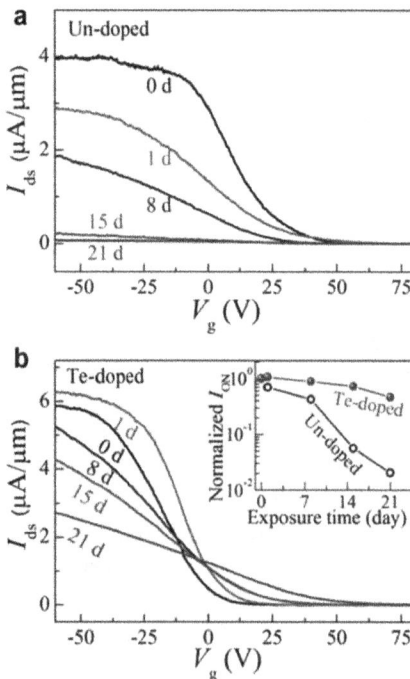

Fig. 4.14: Comparison of undoped and Te-doped BP FETs under ambient conditions.
(a) Transfer curves of an undoped BP device.
(b) Transfer curves of a Te-doped BP device. Inset shows the normalized ON current by the initial value of 0 day and its variation with exposure time (reprinted with permission from John Wiley and Sons Inc, *Adv. Mat.* [79], Copyright John Wiley and Sons Inc. (2019)).

will remain a dream. We have to consider these works mainly for the scientific contribution they brought to the knowledge of the properties of these materials. More consideration will be done at the end of this book.

4.5.1.2 Examples of BP-based transistors: some pioneering works

One of the first pioneering contributions was issued by Li and co-workers in 2014 [80], where they performed the fabrication of p-type FET using few layered BP with a channel thickness of only 7.5 nm. The BP FET showed drain current modulation up to 10^5, and a field-effect mobility value reaching 1,000 cm^2/V s at ambient conditions. The structure of the device is highlighted in Fig. 4.15. However, we have to point out that measures were performed under vacuum, which is not the better conditions to think about a potential implementation in real devices. Anyway, it has to be recognized that this was one of the first works and that the passivation approaches, previously presented, were not optimized yet.

These results were confirmed by the work of Ye and co-workers [91] 2 years later and confirmed the same order in terms of mobility (~300 cm^2/V s) and on/ off ratio (~10^4) (see Fig. 4.16). In the same work, scientists demonstrated the possibility of phosphorene integration by fabricating the first 2D CMOS inverter of phosphorene P-type metal-oxide semiconductor (PMOS) and MoS$_2$ N-type metal-oxide semiconductor (NMOS) transistors [81]. Indeed, they used phosphorene in combination with other 2D materials providing unique 2D heterojunctions. One of the devices was a BP/MoS$_2$ 2D heterojunction as a 2D PN diode, which is one of the basic building blocks for optoelectronic devices. Scientists observed that under the effect of light exposure, the ultra-thin p–n diodes demonstrated a maximum photodetection responsivity of 418 mA/W at the wavelength of 633 nm. Moreover, scientists observed a photovoltaic energy conversion with an external quantum efficiency of 0.3%. Therefore, PN diodes seemed to have interesting properties to target applications in the field of broadband photodetection and solar energy harvesting.

These first very interesting works triggered a lot of other contributions on the effect of contacts and on the optimization of devices [83–85]. Finally, we can mention a recent work where researchers fabricated high-performance devices tested in ambient air. They exploited the super-hydrophobicity featured by fluoroalkylsilane-coated titanium dioxide (TiO$_2$) nanoparticles that were used to passivate the surface of BP FET.

In Fig. 4.17, it is highlighted that the passivation did not change the composition of the layer, thanks to passivation, its properties were preserved. Indeed, after 28 days in ambient conditions (under the effect of oxygen and moisture), the performance presented only 20% of the channel current degradation and the device was effective after 60 days.

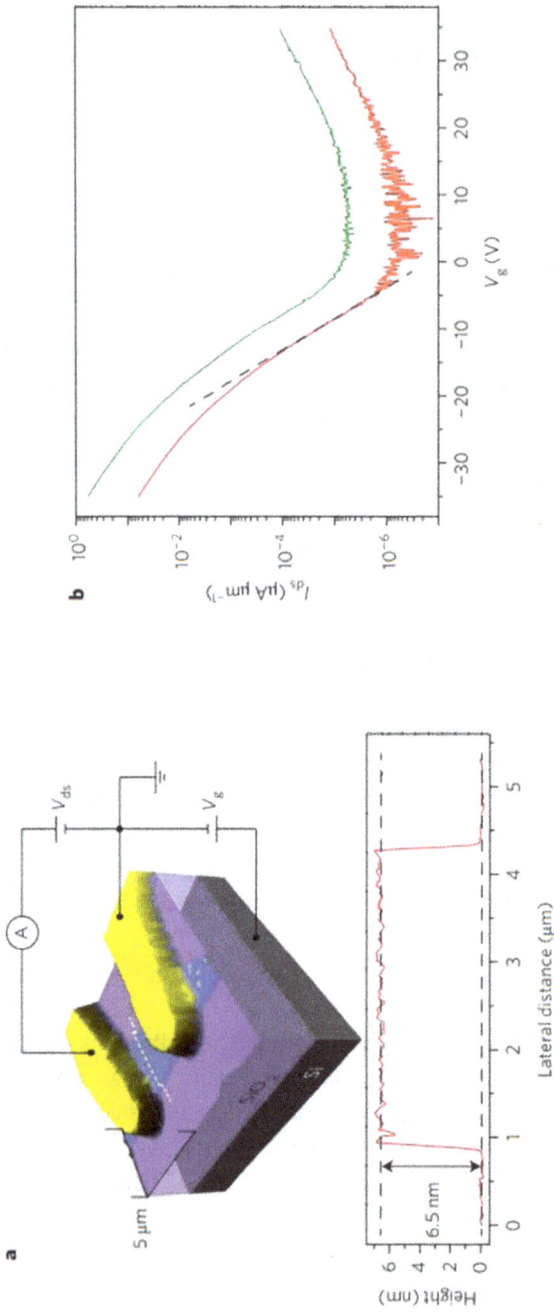

Fig. 4.15: (a) Top: schematic of a device structure of a BP FET. Bottom: cross section of a device along the white dashed line in the schematic. (b) Source–drain current as a function of gate voltage (reprinted with permission from Springer Nature Customer Service Centre GmbH, Springer Nature [80], Copyright Springer (2014)).

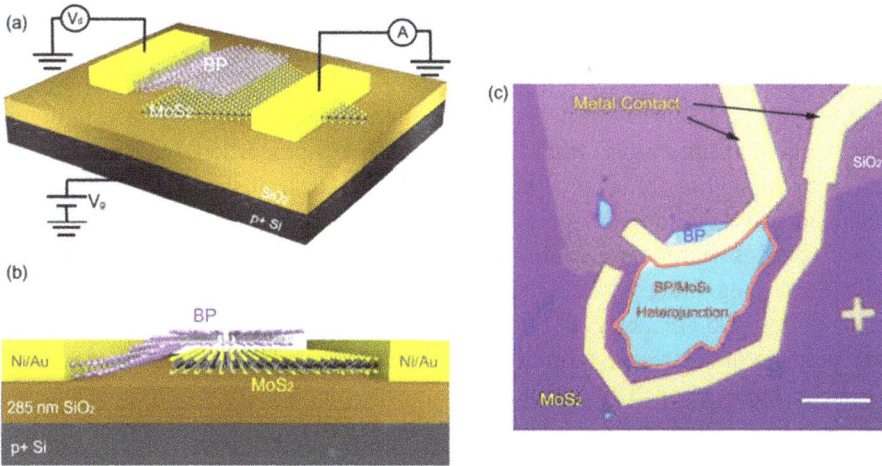

Fig. 4.16: (a) and (b) Schematics of the device structure. Few layered BP flakes exfoliated onto monolayer MoS$_2$ to form a vdW heterojunction. (c) Gate-tunable IV characteristics of the 2D p–n diode [82] (reprinted with permission from [82], Copyright (2014) American Chemical Society).

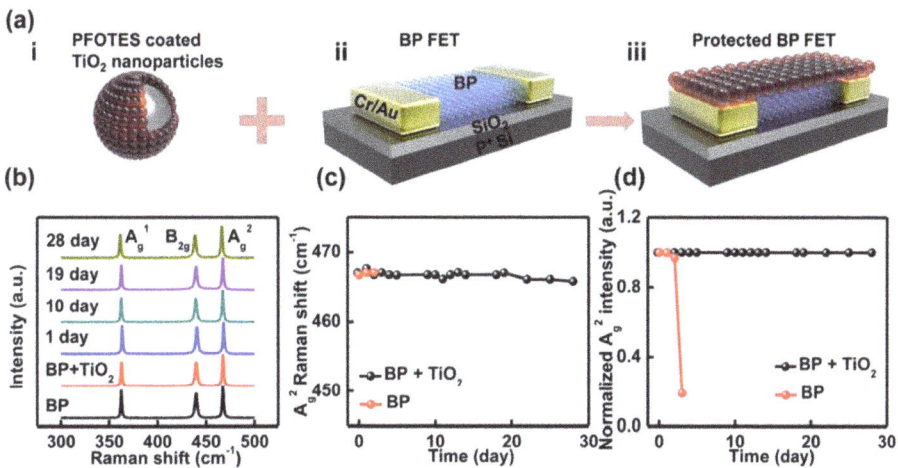

Fig. 4.17: (a) Schematic illustrations of the passivated BP transistor. (b) Raman spectrum of the PFOTES-coated TiO$_2$ nanoparticle-encapsulated BP in ambient conditions. The Raman shift (c) and the normalized peak intensity (d) of A_g^2 at different ageing times (reprinted with permission from [86], Copyright (2019) AIP Publishing).

4.5.2 Energy storage applications

4.5.2.1 Solar cells

As presented in the previous sections, the main feature of BP is the possibility to obtain an ad hoc direct gap, which is simply inversely proportional to the thickness and so to the number of layers that is piled up. Moreover, it shows direct band gap independently from the number of layers, very good mobility and ambipolar transport properties that make us think that it could be an effective material for solar cells. Some pioneering works in 2014 [86] and 2016 [87] seem to confirm that the power conversion efficiency (PCE) could reach a very significant value of 20%. The first example of solar cells fabricated by exploiting few layered BP flakes was reported by Lin et al. in 2019 [88]. In this contribution, the effectiveness of BP was tested in hole transportation layer and in the electron transport layer of organic photovoltaic (OPV) cells. However, the preliminary results did not match the expectations. The main reason evoked by researchers was the mismatched energy alignment of the BP layer in the device, trapping holes and so reducing their charge transport. Researchers observed that the integration of BP flakes onto ZnO electrodes was able to enhance the PCE of OPVs due to the matching band structure of BP between ZnO and PC71BM (which is a derivative of fullerene C70), from 7.37% to 8.25% compared to a cell without BP nanosheets (see Fig. 4.18) [89].

Another important topic of great interest in the field of solar cells is the identification of alternative materials for Pt-coated electrodes for dye-synthesized solar cells (DSSCs) (see Fig. 4.19). Yang and co-workers in 2016 [90] synthesized BP quantum dots (BPQDs) that were mixed with polyaniline film to fabricate electrodes for DSSCs. The extremely interesting results stressed up an enhancement of around 20% using the obtained composite compared to cells without BPQDs. The increased current density increased was mainly associated with the light absorption in the near-infrared region that was possible, thanks to the insertion of BP-based materials.

In the last years, the great tendency in the field of solar cells has been to develop perovskite-based devices. Indeed, the study of perovskite solar cells (PSCs) is a booming field of research, maybe the most popular in the scientific community, thanks to their huge PCE. Moreover, PSCs are less expensive, compared to silicon-based solar cells, and fabrication can be performed by simple wet chemical process. Many research teams focused their work trying to understand the interest in using BP in this kind of devices. A very interesting paper issued by McDonald and co-workers in 2020 [91, 92] shows the utilization of phosphorene nanoribbons, produced by Watts and co-workers [93], in addition to the semiconducting polymer in the perovskite layers. Thanks to that, scientists were able to enhance the extraction of positive charges or holes. Indeed, it seemed that adding phosphorene nanoribbons helps to reach an energy alignment between the layers, achieving a very favourable configuration for hole extraction, also thanks to the very good mobility of charges in phosphorene. Thanks to the integration of phosphorene nanoribbons, the hole transport/

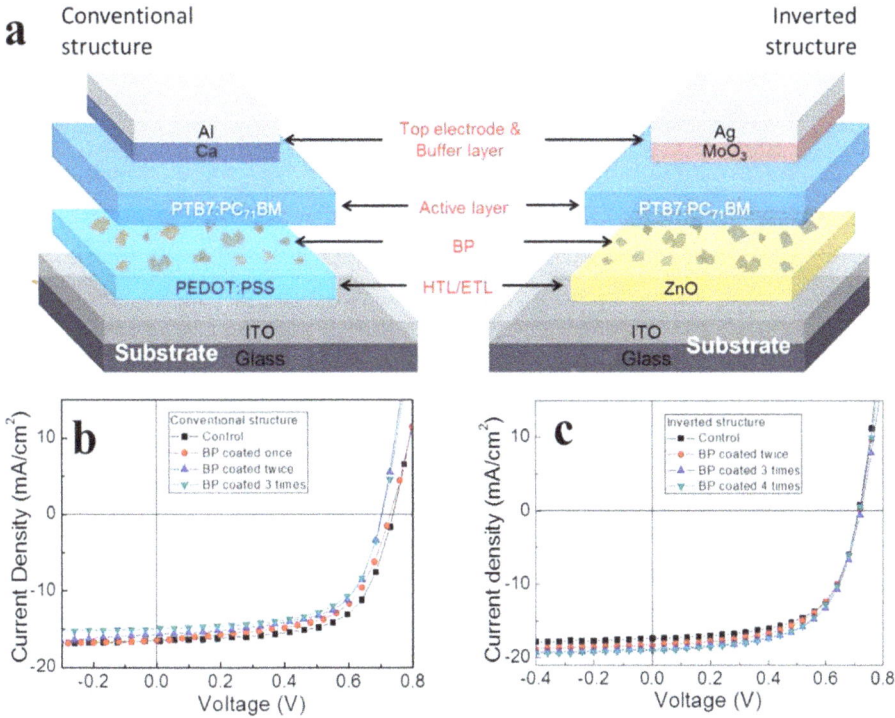

Fig. 4.18: (a) Conventional and inverted architectures of OPVs; *J–V* characteristics of (b) conventional and (c) inverted OPVs with BP incorporation under different conditions (reprinted with permission from [89], Copyright John Wiley and Sons Inc. (2016)).

MAPbI3 interface reached a fill factor above 0.83 with efficiencies exceeding 21% for planar p–i–n (inverted) PSCs. Gong and co-workers used another approach in 2020 [94] fabricating solar cells integrating BPQDs in CsPbI2Br perovskite crystalline thin layers. There were mainly two advantages of this approach where the lone electron pair of BP is strongly bound to CsPbI2Br. Indeed, the BPQDS/CsPbI2Br core–shell structure enhanced the strength leading to a stable CsPbI2Br crystallite and suppresses the oxidation of BPQDs. Therefore, a PCE of 15.47% was reached using 0.7 wt% of BPQDs embedded in CsPbI2Br film-based devices. Another important issue in the case of perovskite-based solar cells is the photostability. In this optics, to enhance this last one, Wand et al. in 2019 [95] incorporated BP into $CH_3NH_3PbI_3$ perovskites (MAPbI$_3$/ BP). The MAPbI$_3$/BP-based PSCs were able to show 94% of the initial efficiency after 1,000 h under exposure of white LED light when the same PSCs without BP decreased their efficiency to reach only 30%. This result underlined that the utilization of material was able to, at the same time, inhibit PbO defect formation and retard charge recombination, such as BP, which is an extremely interesting approach to substantially improve the photostability of organic–inorganic hybrid perovskite-based PSCs. We can conclude that this very brief analysis of some of the main works highlighting the

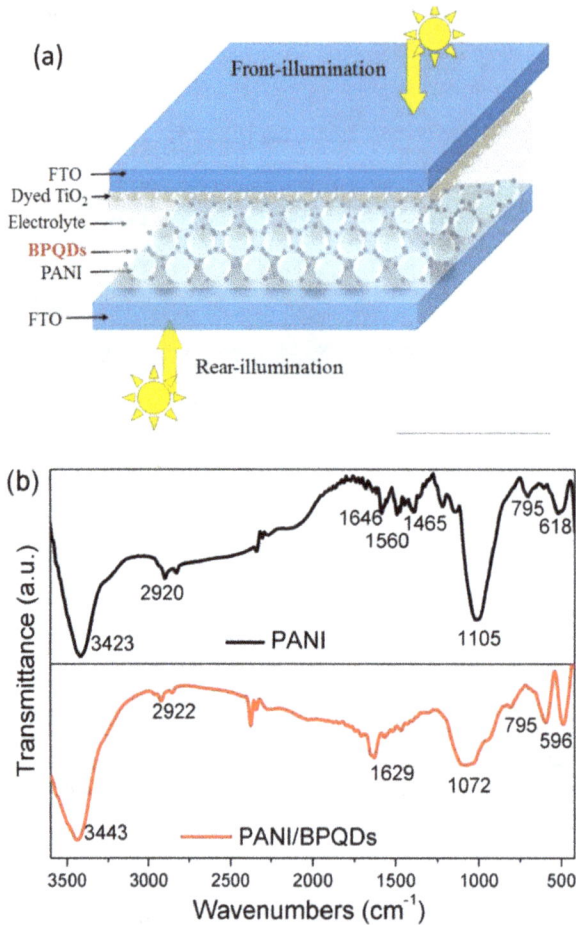

Fig. 4.19: (a) Schema of DSSC with BPQDs–polyaniline-based CE. (b) Comparison between the absorption of a cell with BPQD integration and without (reprinted with permission from [90], Copyright John Wiley and Sons Inc. (2016)).

potential applications of BP for solar cells that the performances reported, especially exploiting BPQDs, seem to be extremely promising and quite easily to be implemented. The only consideration concerns the final cost of the device and on the handling of the nanomaterials. The approach to embed phosphorene in composites or integrated in structures can be very useful and allow achieving the stability of the materials, which is not directly in contact with external moisture or oxygen.

4.5.3 Energy storage applications

Thanks to its advantageous structural and electrochemical properties, such as its large theoretical capacity, high carrier mobility and low redox potential, BP is a potential strong candidate material that can be implemented in the next-generation energy storage devices. The main, and important, drawbacks concern its potential utilization in real device (because of its reactivity with oxygen and moisture), its synthesis, its large volume expansion during cycling and, finally, its poor electronic conductivity. Many efforts have been focused on these issues trying to improve its synthesis methods and electrochemical performance. A large panel of BP-based composite materials have been developed and deeply studied. In this section, we provide a short up-to-date account of the recent progress made in research on BP-based materials for use in lithium-based batteries and supercapacitors.

4.5.3.1 Batteries (lithium ion)
BP, showing a high theoretical specific capacity of 2,596 mAh/g and with large charge-carrier mobility, from the very beginning has presented a great interest for its potential implementation in lithium-ion battery [96]. In one of the first works in the fields, Cui et al. in 2014 [97] applied a strategy for the synthesis of BP–graphite composite anodes exploiting the strong bonds between phosphorus and carbon atoms. This led to a specific capacity of 2,786/2,382 mAh/g during the lithiation/delithiation process, allowing to reach a coulombic efficiency of 85%, compared with only 58% obtained without the use of BP. Some scientists proposed also the utilization of metal-organic frameworks (MOFs) that allowed managing better the expansion of BP layers. In this context, Zhou et al. [98] conceived a 2D BP/NiCo MOF hybrid obtained after sonication steps in a benzenedicarboxylic acid (BDC) solution. The BP layers acted as the scaffold to bind the NiCo MOF. The carboxylate groups in BDC^{2-} chelating with metal ions and bonding with BP led to an increase in the electrochemical performance with good reversible capacity (569 mAh/g at 2A/g after 250 cycles) and very promising rate capability (398 mAh/g at 5 A/g after 1,000 cycles). An interesting contribution by Zhang and co-workers in 2020 [99] proposed the utilization of hybrid synergetic mixture of BP and Carbon nanotubes (CNTs). CNTs indeed strongly enhanced the conductivity creating a sort of conductive backbone, relying also on the BP layers [100]. BP was bound to the CNT matrix that limited significantly its volume expansion during cycling and enhanced transport properties of holes and electrons. It is not the objective of this book to give an exhaustive description of all the approaches but only a global overview. To obtain more details on other battery concepts (e.g. zinc–nickel, sodium based or lithium sulphur), we advise the readers to check for more information in the review paper quoted such as the extremely interesting contribution from Li et al. published in 2021 [104].

4.5.3.2 Supercapacitors

In case of implementation of supercapacitors, one of the main characteristics of 2D materials is the potential implementation in devices integrated on flexible substrates. Indeed, flexible supercapacitors have recently created strong interest, also thanks to their possible applications in the framework of wearable devices. The main hurdles to overcome consist in developing materials able to achieve the necessary performances in terms of energy and power even after being submitted to stretching, compression or torsion. We can mention a very interesting paper by Wu et al. in 2018 [101] (see Fig. 4.20), where scientists developed a hetero-structured material made of BP chemically bound with CNTs, as also achieved in a previous case in batteries. The BP/CNTs composite was integrated in non-woven fibre fabrics to achieve a supercapacitor electrode. The flexible supercapacitor exhibits high energy density (96.5 mWh/cm^3), large volumetric capacitance (308.7 F/cm^3), long cycle stability and durability upon deformation. The enhanced performances were obtained creating a favourable network of pores (increasing the sites that can host charges), improving the conductivity (and so the power delivered) and achieving a mechanical robust structure. This design can implement the utilization of various electronics such as Light-emitting diode (LED), smart watches or flexible displays.

In the same context, another interesting application is for fabrication of high-energy micro-supercapacitors (MSCs). In a very interesting paper, Xiao and co-workers in 2017 [102] developed an easy and rapid process consisting in only one step of mask-assisted fabrication of high-energy MSCs with the interdigital hybrid electrode patterns of stacking high-quality phosphorene nanosheets and electro-chemically exfoliated graphene using IL electrolyte (see Fig. 4.21). The interdigital hybrid electrode films were directly manufactured by layer-by-layer deposition of phosphorene and graphene nanosheets using a customized interdigital mask, and transferring directly onto a flexible substrate.

The resultant patterned interdigital hybrid electrode films present outstanding uniformity, flexibility, conductivity (319 S/cm), and structural integration, which can directly serve as binder- and additive-free flexible electrodes for MSCs. After tests, interdigital hybrid electrode-MSCs delivered impressing energy density of 11.6 mWh/cm^3, larger than most nanocarbon-based MSCs, and showed excellent flexibility and stable performance with very reduced variation of capacitance value also under stretching, compression and torsion. A very important aspect is that the process developed by Xiao et al. is extremely simple and highly versatile for simplified production of parallel and serially interconnected modular power sources.

4.6 Conclusions

In this chapter, we have presented the main physical and chemical properties of BP. We have seen that BP has been "forgotten" for around 100 years. The discovery of the terrific properties of graphene and of the other 2D materials has suddenly changed the

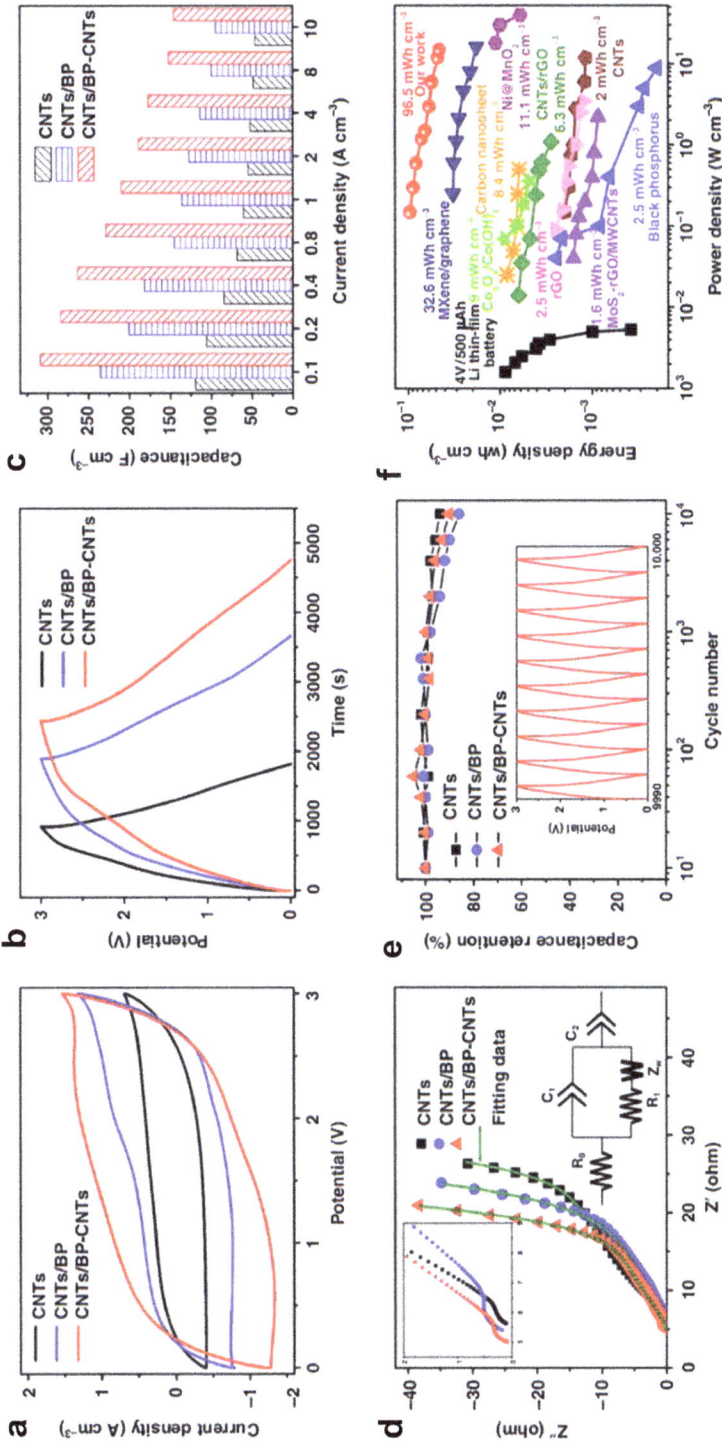

Fig. 4.20: (a) Cyclic voltammetry curves of supercapacitors. (b) Galvanostatic charge/discharge. (c) Calculated specific capacitances under different current densities. (d) Electrochemical impedance spectroscopy (EIS) analysis of SCs. (e) Cyclic testing of SCs under a voltage of 3 V at a current density of 0.4 A/cm³; inset: galvanostatic charge/discharge curves after 10,000 cycles. (f) Energy density versus power density of SCs compared with other electrode-based energy storage systems (reprinted (adapted) with permission from [101]).

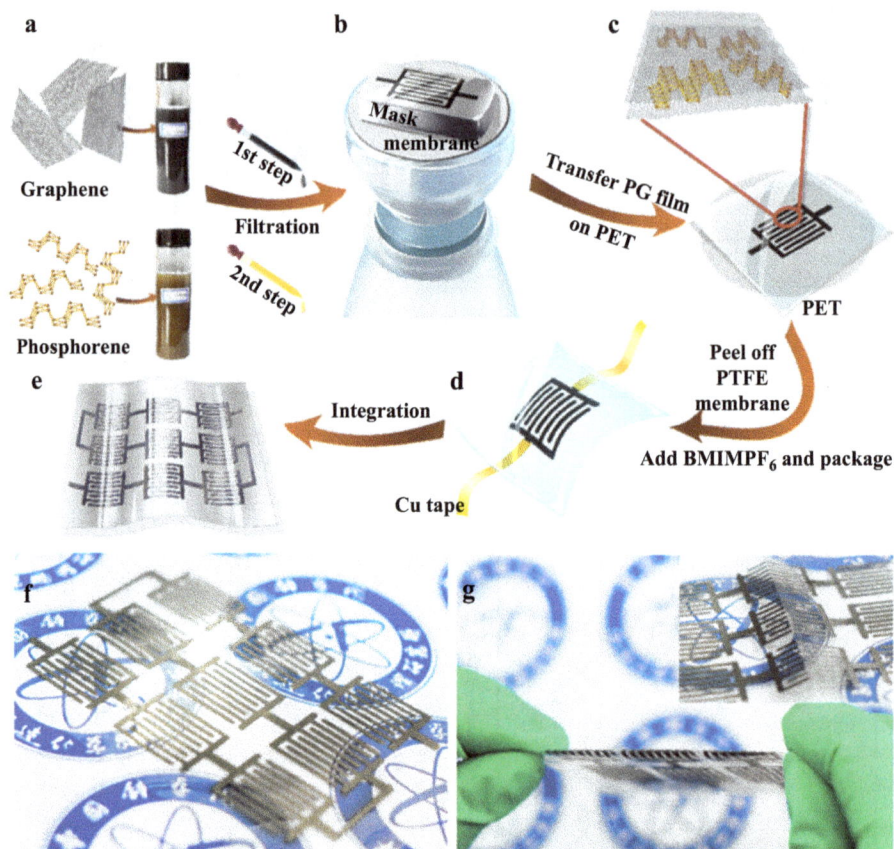

Fig. 4.21: Mask-assisted simplified fabrication of interdigital hybrid electrode MSCs. (a) synthesis of graphene and phosphorene inks; (b) step-by-step filtration of graphene and phosphorene in sequence with the assistance of an interdigital mask; (c) dry transfer of interdigital hybrid electrode film onto PET substrate; (d) peeling off the PTFE membrane; (e) integration of serially interconnected MSC devices; (f) photograph of interconnected interdigital hybrid electrode-MSCs; (g) flexibility and stability demonstration of MSC devices fabricated in flexible surfaces (reproduced with permission [102], Copyright 2017, American Chemical Society).

landscape. Many scientific teams have focused their work on this material in the last 7/8 years with a consequent strong increase in the scientific production. It is true that thanks to the interesting properties in terms of charge mobility and conductivity, some teams tried immediately to implement BP in transistors, as already happened for other 2D and graphene too. The utilization of BP for FET seems to be more curious than a real potential application. The main interest of all the works related to these applications concerns more the scientific investigations of the properties of the materials than the real implementation in industrial transistors. Our objective was to give a quite large overview of the main scientific challenges concerning these materials and its potential implementation in devices. We can observe that to handle this material in ambient conditions is a harsh challenge. Different strategies have been elaborated to passivate BP without changing its intrinsic properties such as encapsulation, functionalization. This will be the main hurdle to overcome in order to target real applications that will be able to strike our everyday life. Some applications such as flexible supercapacitors seem to be extremely interesting. Integration in solar cells is another potential application that has attracted a strong interest also because BP composites can be integrated in perovskite-based solar cells achieving improved performances, especially under the form of BPQDs. In addition, the utilization in lithium-based batteries seems to show promising results. However, we are quite far from the utilization also in these cases in everyday life devices. Another parameter to take into consideration is the possibility to fabricate and exploit the materials developing low-cost processes, which is a strong challenge considering the handling issues. All the applications discussed in this chapter are mainly related to a "more-than-Moore" vision where the materials does not change the present vision of the technology or does not trigger a shift in the paradigm. Indeed BP seems to allow strong improvements in opto-electronics devices thanks to its peculiar characteristics especially its direct band gap which is independent from the number of layer pile-up. Some studies using BP in spintronics are in progress but they are quite limited [103–105] now. We will develop this, maybe, in a new book in the future if major disclosure in the field takes place.

References

[1]   Liu, H., Du, Y. C., Deng, Y. X., Ye, P. D.. Chem. Soc. Rev.. 44, 2015, 2732.
[2]   Zhu, X. J., Zhang, T. M., Sun, Z. J., Chen, H. L., Guan, J., Chen, X., Ji, H. X., Du, P. W., Yang, S. F.. Adv. Mater. 29, 2017, 1605776.
[3]   Chowdhury, C., Datta, A. J. Phys. Chem. Lett. 8, 2017, 2909.
[4]   doi:10.1038/natrevmats.2016.61
[5]   https://www.npr.org/sections/health-shots/2016/02/02/465188104/phosphorus-starts-with-pee-in-this-tale-of-scientific-serendipity?t=1638309677883
[6]   Two new modifications of phosphorous. Bridgman, P. W. J. Am. Chem. Soc. 36, 7, 1914, 1344–1363.
[7]   Zhu, Z., Tománek, D. Semiconducting layered blue phosphorus: A computational study. Phys. Rev. Lett. 112, 2014, 176802.
[8]   Guo, H., Lu, N., Dai, J., Wu, X., Zeng, X. C. Phosphorene nanoribbons, phosphorus nanotubes, and van der Waals multilayers. J. Phys. Chem. C. 118, 2014, 14051–14059.
[9]   Guan, J., Zhu, Z., Tománek, D. Phase coexistence and metal–insulator transition in few-layer phosphorene: A computational study. Phys. Rev. Lett. 113, 2014, 046804.
[10]   Wittig, J., Matthias, B. T. Superconducting phosphorus. Science. 160, 1968, 994–995.
[11]   Rajagopalan, M., Alouani, M., Christensen, N. Calculation of band structure and superconductivity in the simple cubic phase of phosphorus. J. Low Temp. Phys. 75, 1989, 1–13.

[12] Chan, K. T., Malone, B. D., Cohen, M. L. Pressure dependence of superconductivity in simple cubic phosphorus. Phys. Rev. B. 88, 2013, 064517.
[13] Karuzawa, M., Ishizuka, M., Endo, S. The pressure effect on the superconducting transition temperature of black phosphorus. J. Phys. Condens. Matter. 14, 2002, 10759.
[14] https://www.nobelprize.org/prizes/physics/1946/summary/
[15] www.pnas.org/cgi/doi/10.1073/pnas.1416581112
[16] Jamieson, J. C. Crystal structures adopted by black phosphorus at high pressures. Science. 139, 3561, 1963, 1291–1292.
[17] Brown, A., Rundqvist, S. Refinement of the crystal structure of black phosphorus. Acta Crystallogr. 19, 4, 1965, 684–685.
[18] Akahama, Y., Endo, S., Narita, S. Electrical properties of black phosphorus single crystals. J. Phys. Soc. Jpn. 52, 6, 1983, 2148–2155.
[19] Ikezawa, M., Kondo, Y., Shirotani, I. Infrared optical absorption due to one and two phonon processes in black phosphorus. J. Phys. Soc. Jpn. 52, 5, 1983, 1518–1520.
[20] Sugai, S., Shirotani, I. Raman and infrared reflection spectroscopy in black phosphorus. Solid State Commun. 53, 9, 1985, 753–755.
[21] Shibata, K., Sasaki, T., Morita, A. The energy band structure of black phosphorus and angle-resolved ultraviolet photoelectron spectra. J. Phys. Soc. Jpn. 56, 6, 1987, 1928–1931.
[22] Narita, S. et al., Far-infrared cyclotron resonance absorptions in black phosphorus single crystals. J. Phys. Soc. Jpn. 52, 10, 1983, 3544–3553.
[23] Keyes, R. The electrical properties of black phosphorus. Phys. Rev. 92, 3, 1953, 580–584.
[24] Morita, A., Sasaki, T. Electron-phonon interaction and anisotropic mobility in black phosphorus. J. Phys. Soc. Jpn. 58, 5, 1989, 1694–1704.
[25] Suzuki, N., Aoki, M. Interplanar forces of black phosphorus caused by electron-lattice interaction. Solid State Commun. 61, 10, 1987, 595–600.
[26] Asahina, H., Morita, A. Band structure and optical properties of black phosphorus. J. Phys. C Solid State Phys. 17, 11, 1984, 1839–1852.
[27] Kawamura, H., Shirotani, I., Tachikawa, K. Anomalous superconductivity and pressure induced phase transitions in black phosphorus. Solid State Commun. 54, 9, 1985, 775–778.
[28] Bondavalli, P. Graphene and related nanomaterials 1st edn Properties and Applications. Hardcover. 2017. Amsterdam, Elsevier, 192.
[29] https://www.nobelprize.org/prizes/physics/2010/summary/
[30] Morita, A. Semiconducting black phosphorus. Appl. Phys. A. 39, 1986, 227–242, https://doi.org/10.1007/BF00617267.
[31] Scelta, D., Baldassarre, A., Serrano-Ruiz, M., Dziubek, K., Cairns, A. B., Peruzzini, M., Bini, R., Ceppatelli, M. Interlayer Bond Formation in Black Phosphorus at High Pressure. Angew. Chem. Int. Ed. Engl. 2017 Nov 6 56, 45, 14135–14140. 10.1002/anie.201708368.
[32] Suzuki, N., Aoki, M. Interplanar forces of black phosphorus caused by electron-lattice interaction. Solid State Commun. 61, 10, 1987, 595–600.
[33] Asahina, H., Shindo, K., Morita, A. Electronic structure of black phosphorus in self-consistent pseudopotential approach. J. Phys. Soc. Jpn. 51, 4, 1982, 1193–1199. https://doi.org/10.1143/JPSJ.51.1193.
[34] Takao, Y., Morita, A. Electronic structure of black phosphorus: Tight binding approach. Physica B+C. 105, Issues 1–3, 1981, https://doi.org/10.1016/0378-4363(81). 90222–90229.
[35] Ling, X., Wang, H., Huang, S., Xia, F., MS, D. The renaissance of black phosphorus. Proc. Natl. Acad. Sci. U.S.A. 2015 Apr 14 112, 15, 4523–4530. 10.1073/pnas.1416581112.
[36] 2D Black Phosphorus: From Preparation to Applications for Electrochemical Energy Storage, Wu, S., Hui, K. S., Hui, K. N., 5, 5, 2018; https://onlinelibrary.wiley.com/doi/10.1002/advs.201700491

[37] Sun, J., Zheng, G. Y., Lee, H. W., Liu, N., Wang, H. T., Yao, H. B., Yang, W. S., Cui, Y. Nano Lett. 14, 2014, 4573.

[38] Appalakondaiah, S., Vaitheeswaran, G., Lebegue, S., Christensen, N. E., Svane, A. Phys. Rev. B. 86, 2012, 035105.

[39] Iwasaki, H., Kikegawa, T., Fujimura, T., Endo, S., Akahama, Y., Akai, T., Shimomura, O., Yamaoka, S., Yagi, T., Akimoto, S., Shirotani, I. Physica B+C. 139, 1986, 301.

[40] Aldave, S. H., Yogeesh, M. N., Zhu, W. N., Kim, J., Sonde, S. S., Nayak, A. P., Akinwande, D. 2D Mater. 3, 2016, 0144007.

[41] Morita, A. Semiconducting black phosphorus. Appl. Phys. A. 39, 1986, 227–242, https://doi.org/10.1007/BF00617267.

[42] Ahuja, R. Phys. Status Solidi B. 235, 2003, 282. https://doi.org/10.1002/pssb.200301569.

[43] Li, L., Yu, Y., Ye, G. et al., Black phosphorus field-effect transistors. Nature Nanotech. 9, 2014, 372–377. https://doi.org/10.1038/nnano.2014.35.

[44] Liu, H., Neal, A. T., Zhu, Z., Luo, Z., Xu, X. F., Tomanek, D., Ye, P. D. ACS Nano. 8, 2014, 4033. https://doi.org/10.1021/nn501226z.

[45] Tran, V., Soklaski, R., Liang, Y., Yang, L. Layer controlled band gap and anisotropic excitons in fewlayer black phosphorus. Phys. Rev. B. 89, 2014, 235319.

[46] Rahman, M. Z., Kwong, C. W., Davey, K., Qiao, S. Z. Energy Environ. Sci. 9, 2016, 709.

[47] Carvalho, A., Wang, M., Zhu, X. et al., Phosphorene: From theory to applications. Nat. Rev. Mater. 1, 2016, 16061. https://doi.org/10.1038/natrevmats.2016.61.

[48] Favron, A., Gaufrès, E., Fossard, F., Phaneuf-L'Heureux, A. L., Tang, N. Y., Lévesque, P. L., Loiseau, A., Leonelli, R., Francoeur, S., Martel, R. Photooxidation and quantum confinement effects in exfoliated black phosphorus. Nat Mater. 14, 8, 2015, 826–832. 10.1038/nmat4299.

[49] Castellanos-Gomez, A., Vicarelli, L., Prada, E., Island, J. O., Narasimha-Acharya, K. L., Blanter, S. I., Groenendijk, D. J., Buscema, M., Steele, G. A., Alvarez, J. V., Zandbergen, H. W., Palacios, J. J., Herre, S. J. V. D. Z. 2D Mater. 1. 2014, 02500. https://doi.org/10.1088/2053-1583/1/2/025001.

[50] Du, Y., Ouyang, C., Shi, S., Lei, M. Ab initio studies on atomic and electronic structures of black phosphorus. J Appl. Phys. 107, 2010, 09D908. https://doi.org/10.1063/1.3386509.

[51] Effective Passivation of Black Phosphorus under Ambient Conditions, Jongchan Yoon, and Zonghoon Lee. Applied Microscopy. 47, 2017, 176–186. 2017. https://doi.org/10.9729/AM.2017.47.3.176.

[52] Island, J. O., Steele, G. A., van der Zant, H. S. J., Castellanos-Gomez, A. 2D Mater. 2, 2015, 011002. https://doi.org/10.1088/2053-1583/2/1/011002.

[53] Zhou, Q., Chen, Q., Tong, Y., Wang, J. Light-Induced Ambient Degradation of Few-Layer Black Phosphorus: Mechanism and Protection. Angew. Chem. Int. Ed. Engl. 2016 Sep 12, 55, 38, 11437–11441. 10.1002/anie.201605168.

[54] Effective Passivation of Exfoliated Black Phosphorus Transistors against Ambient Degradation. Wood, J. D., Wells, S. A., Jariwala, D., Chen, K.-S., Cho, E., Sangwan, V. K., Liu, X., Lauhon, L. J., Marks, T. J., Hersam, M. C. Nano Lett. 14, 12, 2014, 6964–6970, 20140. https://doi.org/10.1021/nl5032293.

[55] Illarionov, Y. Y., Waltl, M., Rzepa, G. et al., Highly-stable black phosphorus field-effect transistors with low density of oxide traps. Npj 2D Mater Appl. 1, 23, 2017, https://doi.org/10.1038/s41699-017-0025-3.

[56] Galceran, R., Gaufres, E., Loiseau, A., Piquemal-Banci, M., Godel, F., Vecchiola, A., Bezencenet, O., Martin, M.-B., Servet, B., Petroff, F., Dlubak, B., Seneor, P. Stabilizing ultra-thin black phosphorus with in-situ-grown 1 nm-Al2O3 barrier. Appl. Phys. Lett. 111, 2017, 243101, https://doi.org/10.1063/1.5008484.

[57]   Kern, L. M., Galceran, R., Zatko, V., Galbiati, M., Godel, F., Perconte, D., Bouamrane, F.,
       Gaufrès, E., Loiseau, A., Brus, P., Bezencenet, O., Martin, M.-B., Servet, B., Petroff, F.,
       Dlubak, B., Seneor, P. Atomic layer deposition of a MgO barrier for a passivated black
       phosphorus spintronics platform, L.-M. Kern. Appl. Phys. Lett. 114, 2019, 053107, https://
       doi.org/10.1063/1.5086840.
[58]   Sun, J., Choi, Y., Choi, Y. J. et al., Adv. Mater. 31, 2019, 1803831. https://doi.org/10.1002/
       adma.201803831.
[59]   Bertolazzi, S., Gobbi, M., Zhao, Y. et al., Chem. Soc. Rev. 47, 2018, 6845–6888. https://doi.
       org/10.1039/C8CS00169C.
[60]   Huang, Y. L., Zheng, Y. J., Song, Z. et al., Chem. Soc. Rev. 47, 2018, 3241–3264. https://doi.
       org/10.1039/C8CS00159F.
[61]   Azadmanjiri, J., Kumar, P., Srivastava, V. K. et al., ACS Appl. Nano Mater. 3, 2020, 3116–3143.
       https://doi.org/10.1021/acsanm.0c00120.
[62]   Yang, Y. S., Yang, X. G., Fang, X. Y. et al., Adv. Sci. 5, 2018, 1801187. https://doi.org/
       10.1002/advs.201801187.
[63]   Zhou, B., Yan, D. P. Angew. Chem. Int. Ed. 58, 2019, 15128. https://doi.org/10.1002/
       anie.201909760.
[64]   Sang, D. K., Wang, H., Guo, Z. et al., Adv. Funct. Mater. 29, 2019, 1903419. https://doi.org/
       10.1002/adfm.201903419.
[65]   Gao, J., Zhang, G., Zhang, Y.-W. Nanoscale. 9, 2017, 4219. https://doi.org/10.1039/
       C7NR00894E.
[66]   Sang, D. K., Wang, H., Guo, Z. et al., Adv. Funct. Mater. 29, 2019, 1903419. https://doi.org/
       10.1002/adfm.201903419.
[67]   Ryder, C. R., Wood, J. D., Wells, S. A., Yang, Y., Jariwala, D., Marks, T. J., Schatz, G. C.,
       Hersam, M. C. Covalent functionalization and passivation of exfoliated black phosphorus via
       aryl diazonium chemistry. Nat Chem. 2016 Jun 8, 6, 597–602. 10.1038/nchem.2505.
[68]   Lloret, V., Nuin, E., Kohring, M., Wild, S., Löffler, M., Neiss, C., Krieger, M., Hauke, F.,
       Andreas Görling, H. B., Weber, G. A., Hirsch, A., Noncovalent functionalization and
       passivation of black phosphorus with optimized perylene diimides for hybrid field effect
       transistors, 7, 23, 2020, https://doi.org/10.1002/admi.202001290
[69]   Guo, R., Zheng, Y., Zhirui, M., Lian, X., Sun, H., Han, C., Ding, H., Xu, Q., Yu, X., Zhu, J., Chen,
       W. Surface passivation of black phosphorus via van der Waals stacked PTCDA. Appl. Surf.
       Sci. 496, 2019, 143688, https://doi.org/10.1016/j.apsusc.2019.143688.
[70]   Lei, S. Y., Shen, H. Y., Sun, Y. Y., Wan, N., Yu, H., Zhang, S. Enhancing the ambient stability of
       few-layer black phosphorus by surface modification. RSC Adv. 8, 2018, 14676. https://doi.
       org/10.1039/C8RA00560E.
[71]   Guo, Z., Chen, S., Wang, Z., Yang, Z., Liu, F., Xu, Y., Wang, J., Yi, Y., Zhang, H., Liao, L., Chu,
       P. K., Yu, X. F. Adv. Mater. 29, 2017, 1703811. https://doi.org/10.1002/adma.201703811.
[72]   Zhao, W., Xue, Z., Wang, J., Jiang, J., Zhao, X., Mu, T. ACS Appl. Mater. Interfaces. 7, 2015,
       27608.
[73]   Luo, S., Zhao, J., Zou, J., He, Z., Xu, C., Liu, F., Huang, Y., Dong, L., Wang, L., Zhang, H. ACS
       Appl. Mater. Interfaces. 10, 2018, 3538.
[74]   Zhao, W., Xue, Z., Wang, J., Jiang, J., Zhao, X., Mu, T. ACS Appl. Mater. Interfaces. 7, 2015,
       27608.
[75]   Tang, X., Liang, W., Zhao, J., Li, Z., Qiu, M., Fan, T., Luo, C. S., Zhou, Y., Li, Y., Guo, Z., Zhang,
       H. Small. 13, 2017, 1702739.
[76]   Coiai, S., Cicogna, F., Scittarelli, D., Legnaioli, S., Borsacchi, S., Ienco, A., Serrano-Ruiz, M.,
       Caporali, M., Peruzzini, M., Dinelli, F., Ishak, R., Signori, F., Toffanin, S., Bolognesi, M.,
       Prescimone, F., Passaglia, E. Incorporation of 2D black phosphorus (2D-bP) in P3HT/PMMA

mixtures for novel materials with tuned spectroscopic, morphological and electric features. FlatChem. 30, 2021, 100314, https://doi.org/10.1016/j.flatc.2021.100314.

[77] On the stability of surfactant-stabilised few-layer black phosphorus in aqueous media. Brent, J. R., Ganguli, A. K., Kumar, V., Lewis, D. J., McNaughter, P. D., O'Brien, P., Sabherwal, P., Tedstone, A. A. RSC Adv. 6, 2016, 86955–86958. d DOI. 10.1039/C6RA21296D.

[78] Wang, Z., Lu, J., Wang, J., Li, J., Du, Z., Wu, H., Liao, L., Chu, P. K., Yu, X. F. Air-stable n-doped black phosphorus transistor by thermal deposition of metal adatoms. Nanotechnology. 2019 Mar 29 30, 13, 135201. 10.1088/1361-6528/aafd68.

[79] Yang, B. C., Wan, B. S., Zhou, Q. H., Wang, Y., Hu, W. T., Lv, W. M., Chen, Q., Zeng, Z. M., Wen, F. S., Xiang, J. Y., Yuan, S. J., Wang, J. L., Zhang, B. S., Wang, W. H., Zhang, J. Y., Xu, B., Zhao, Z. S., Tian, Y. J., Liu, Z. Y. Adv. Mater. 28, 2016, 9408 Te-Doped Black Phosphorus Field-Effect Transistors, https://doi.org/10.1002/adma.201603723.

[80] Li, L., Yu, Y., Ye, G. et al., Black phosphorus field-effect transistors. Nature Nanotech. 9, 2014, 372–377. https://doi.org/10.1038/nnano.2014.35.

[81] Liu, H., Neal, A. T., Zhu, Z., Luo, Z., Xu, X., Tomanek, D., Ye, P. D. ACS Nano. 8, 2014, 4033, org/10.1021/nn501226z.

[82] Deng, Y., Zhe Luo, N. J., Conrad, H. L., Gong, Y., Sina Najmaei, P. M., Ajayan, J. L., Xianfan, X., Ye, P. D. Black phosphorus monolayer MoS2 van der Waals heterojunction pn diode. ACS Nano. 8, 8, 2014, 8292–8299. 10.1021/nn5027388.

[83] Liu, X., Ang, K. W., Yu, W. et al., Black phosphorus based field effect transistors with simultaneously achieved near ideal subthreshold swing and high hole mobility at room temperature. Sci Rep. 6, 2016, 24920. https://doi.org/10.1038/srep24920.

[84] Tian, H. et al., Negative capacitance black phosphorus transistors with low SS. IEEE Trans. Electron Devices. 66, 3, 2019, 1579–1583, 10.1109/TED.2018.2890576.

[85] Yang, S., Zhang, K., Ricciardulli, A. G., Zhang, P., Liao, Z., Lohe, M. R., Zschech, E., Blom, P. W. M., Pisula, W., Müllen, K., Feng, X., A delamination strategy for thinly layered defect-free high-mobility black phosphorus flakes. Angewandte Chemie. 130, 17, 2018, https://doi.org/10.1002/ange.201801265.

[86] Dai, J., Zeng, X. C. Bilayer phosphorene: Effect of stacking order on bandgap and its potential applications in thin-film solar cells. J. Phys. Chem. Lett. 5, 2014, 1289.

[87] Hu, W., Lin, L., Yang, C., Dai, J., Yang, J. Edge-modified phosphorene nanoflake heterojunctions as highly efficient solar cells. Nano Lett. 16, 2016, 1675.

[88] Lin, S., Yanyong, L., Qian, J., Lau, S. P. Emerging opportunities for black phosphorus in energy applications. Mater. Today Energy. 12, 2019, 1–25, https://doi.org/10.1016/j.mtener.2018.12.004.

[89] Lin, S., Liu, S., Yang, Z., Li, Y., Ng, T. W., Xu, Z., Bao, Q. Solution-processable ultrathin black phosphorus as an effective electron transport layer in organic photovoltaics. Adv. Funct. Mat. 26, 6, 2016, 864–871. https://doi.org/10.1002/adfm.201503273.

[90] Yang, Y., Gao, J., Zhang, Z., Xiao, S., Xie, H. H., Sun, Z. B., Wang, J. H., Zhou, C. H., Wang, Y. W., Guo, X. Y., Chu, P. K., Yu, X. F. Black Phosphorus Based Photocathodes in Wideband Bifacial Dye-Sensitized Solar Cells. Adv. Mater. Weinheim. 2016 Oct 28, 40, 8937–8944. 10.1002/adma.201602382.

[91] Macdonald, T. J., Clancy, A. J., Xu, W., Jiang, Z., Lin, C. T., Mohan, L., Du, T., Tune, D. D., Lanzetta, L., Min, G., Webb, T., Ashoka, A., Pandya, R., Tileli, V., McLachlan, M. A., Durrant, J. R., Haque, S. A., Howard, C. A. Phosphorene Nanoribbon-Augmented Optoelectronics for Enhanced Hole Extraction. J. Am. Chem. Soc. 2021 Dec 29 143, 51, 21549–21559. 10.1021/jacs.1c08905.

[92] https://www.chemistryworld.com/news/phosphorene-nanoribbons-find-their-first-use-in-a-solar-cell-just-3-years-after-discovery/4015062.article

[93] Watts, M. C., Picco, L., Russell-Pavier, F. S. et al., Production of phosphorene nanoribbons. Nature. 568, 2019, 216–220. https://doi.org/10.1038/s41586-019-1074-x.

[94] Gong, X., Guan, L., Li, Q., Li, Y., Zhang, T., Pan, H., Sun, Q., Shen, Y., Grätzel, C., Zakeeruddin, S. M., Grätzel, M., Wang, M. Black phosphorus quantum dots in inorganic perovskite thin films for efficient photovoltaic application. Sci. Adv. 2020, Apr 10 6, 15, eaay5661. doi, 10.1126/sciadv.aay5661.

[95] Haijuan Zhang, W., Zhang, T., Shi, W., Kan, M., Chen, J., Zhao, Y., Photostability of MAPbI3 perovskite solar cells by incorporating black phosphorus Yong, 3, 9, Special Issue: Perovskite solar cells and optoelectronics Part 1 (2019) https://doi.org/10.1002/solr.201900197

[96] Beladi-Mousavi, S. M., Pumera, M. Chem. Soc. Rev. 47, 2018, 6964.

[97] Sun, J., Zheng, G. Y., Lee, H. W., Liu, N., Wang, H. T., Yao, H. B., Yang, W. S., Cui, Y. Nano Lett. 14, 2014, 4573.

[98] Jin, J., Zheng, Y., Huang, S. Z., Sun, P. P., Srikanth, N., Kong, L. B., Yan, Q. Y., Zhou, K. J. Mater. Chem. A. 7, 2019, 783.

[99] Zhang, Y. P., Wang, L. L., Xu, H., Cao, J. M., Chen, D., Han, W. Adv. Funct. Mater. 30, 2020, 1909372.

[100] The development, application, and performance of black phosphorus in energy storage and conversion. Peng, L., Jianguo, L., Cui, H., Shuangchen, R. A., Zeng, Y.-J. 10.1039/D0MA01016B. (Review Article) Mater. Adv. 2, 2021, 2483–2509.

[101] Wu, X., Xu, Y., Hu, Y. et al., Microfluidic-spinning construction of black-phosphorus-hybrid microfibres for non-woven fabrics toward a high energy density flexible supercapacitor. Nat. Commun. 9, 2018, 4573. https://doi.org/10.1038/s41467-018-06914-7.

[102] One-Step Device Fabrication of Phosphorene and Graphene Interdigital Micro-Supercapacitors with High Energy Density. Xiao, H., Wu, Z.-S., Chen, L., Zhou, F., Zheng, S., Ren, W., Cheng, H.-M., Bao, X. ACS Nano. 11, 7, 2017, 7284–7292. 2017. https://doi.org/10.1021/acsnano.7b03288.

[103] Avsar, A., Tan, J., Kurpas, M. et al., Gate-tunable black phosphorus spin valve with nanosecond spin lifetimes. Nature Phys. 13, 2017, 888–893. https://doi.org/10.1038/nphys4141.

[104] Zhang, L., Chen, J., Zheng, X., Wang, B., Zhang, L., Xiao, L., Jia, S. Gate-tunable large spin polarization in a few-layer black phosphorus-based spintronic device. Nanoscale. 2019 Jun 20 11, 24, 11872–11878. 10.1039/c9nr03262b.

[105] Chen, H., Li, B., Yang, J. Proximity Effect Induced Spin Injection in Phosphorene on Magnetic Insulator. ACS Appl. Mater. Interfaces. 2017 Nov 8 9, 44, 38999–39010. 10.1021/acsami.7b11454.

5 Straintronics: a new way to engineer the physical intrinsic properties of 2D materials and van der Waals structures

5.1 Introduction

Straintronics can be defined as the technique that changes the physical properties of materials (e.g. electronic structure of materials and also their topological properties), applying a strain of different origins on them. This technology has already been employed, especially in semiconductor industry, exploiting the mismatch between different materials or films. Indeed, straintronics could be an extremely efficient strategy to handle the properties of 2D materials integrated in the van der Waals (vdW) architecture. In this case, we are talking about different 2D materials piled up, kept together only by weak interlayer forces, which have been highlighted by the seminal paper of Geim et al. in 2013 [2]. To be clearer, vdW forces, named after the Dutch physicist Johannes Diderik van der Waals, are basically forces depending on the distance between atoms or molecules in different materials whose surfaces are put in contact. Therefore, vdW forces are weak forces that do not have the strength of ionic or covalent bonds that are achieved through the chemical bond [1]. Considering the weakness of these forces, its easy to understand that thanks to strain we can "play" with the properties of the structures and changing them. We know that vdW forces quickly vanish at longer distances between interacting molecules. Exploiting these forces, Geim suggested that the isolated atomic foils can also be put together to design heterostructures made ad hoc layer by layer. Thanks to this, the properties of a vertical heterostructure composed of stacked 2D materials can be tailored simply using piling up different 2D materials (see Fig. 5.1). Therefore, in this "vision" (also called "Lego view" [2], where each brick is a different layer with different physical properties), we can imagine to implement complex functionalities by simply choosing 2D materials with specific properties and different origins.

We have already told that strain engineering is not a new complete strategy that can be applied to change intrinsic properties of materials, especially semiconducting ones. Indeed, it has largely employed in industry during the last few decades. For example, uniaxial or biaxial tensile strain has been applied to the channel of a silicon transistor, to enhance the mobility of charges [3]. The main limit is related to the fact that the single bulk crystal is not able to afford strong strain, and is mainly because of its intrinsic 3D nature. For all these reasons, compared with bulk materials, 2D materials that are able to endure strong deformations, especially in terms of stretching and compression avoiding fracture, show great potentials in strain engineering. In this chapter, we will also focus a part of our contribution on 2D materials and how to implement them in vdW structures applying specific strain to achieve/

https://doi.org/10.1515/9783110656336-006

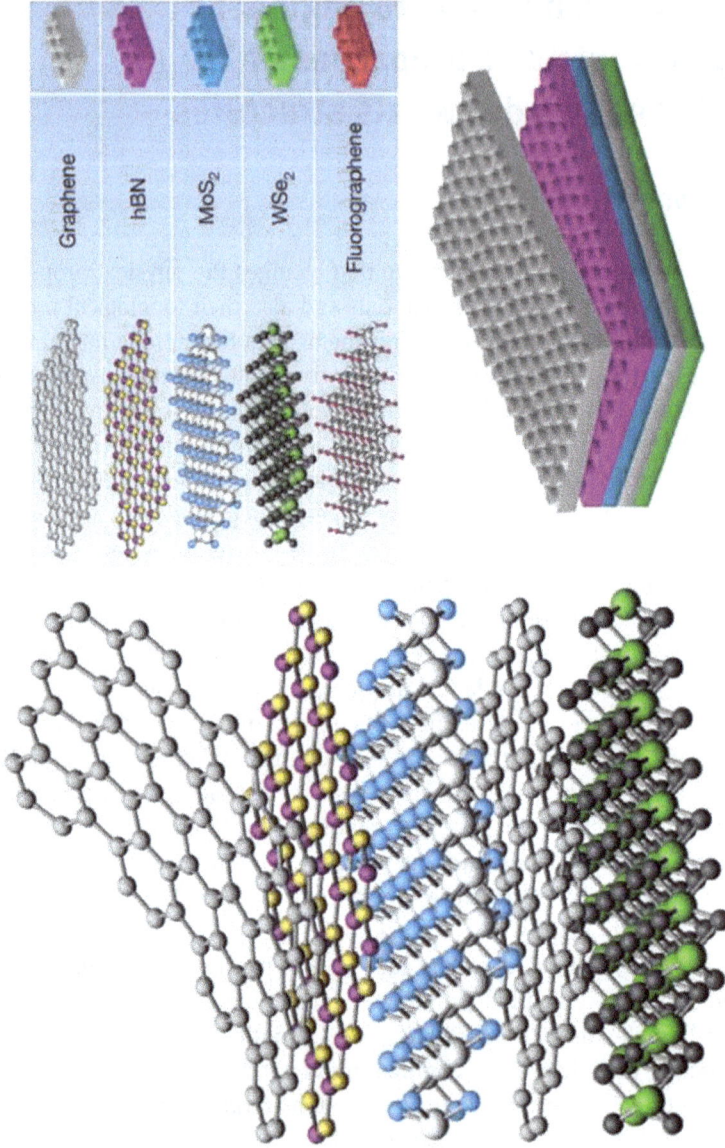

Fig. 5.1: "Lego's view" through vdW heterostructures of 2D materials (reprinted with permission from [3], Copyright © 2013, Nature Springer Publishing Group).

modulate specific properties. To induce the strain, considering the atomic-scale thickness, 2D materials tend to naturally deform in the out-of-plane direction after applying the external strain stimulus. For this reason, one of the fundamental points is the choice of a specific flexible substrate where the 2D layer/layers can be stack. Different kinds of flexible substrates have been used for straintronic implementation such as polydimethylsiloxane (PDMS), polycarbonate, polystyrene, polyethylene terephthalate, polyethylene naphthalate and polyimide [4]. In this chapter, we will start classifying and presenting the different kinds of strains that can be applied and will show some of the most interesting examples that we can find in the literature. The objective of this chapter is to show a global view of the possibilities that can be achieved using this technology to change in an unusual way the properties for 2D materials.

5.2 What type of strain and what are the main effects?

We have decided to focus our attention on 2D materials and vdW structures; as a consequence, following the classification provided by Miao et al. [4] in 2021, we can classify the two main types of strain such as in plane and out of plane. In the first case, the strain is applied by stretching or compressing the 2D materials in a direction parallel to the layers (see Fig. 5.2a). In the second case, the strain is applied perpendicularly to the layers (see Fig. 5.2b).

In the case of out-of-plain strain, we can observe that this strain will have components related to in-plane strain as it is inversely true. For example, we can see that the in-plane strain can be created by applying techniques that create vertical (perpendicular, if reported to the parallel direction of the layers) displacement of layers. For all these reasons, the differences between these techniques are in some case blurred. Concerning heterostrain, we can notice that in the traditional semiconductor industry, strain is achieved in bulk single crystals capitalizing on the mismatch of lattice constants between single-crystal materials and substrates used during the epitaxial growth. This is commonly called heterostrain, and it could be a compressive or tensile type of strain depending on the lattice of the atoms. The method of lattice mismatch can be translated to the strain applied on 2D materials with atomic thickness. When there is a lattice mismatch between two components of 2D heterojunctions or between 2D materials and substrates, tensile or compressive strain is induced in 2D materials [5]. In this chapter, up to now we have considered simply the strain on layers with the same origin. In case of vdW heterostructures using different layers, we can also take into account the heterostrain. In case of 2D materials, it allows inducing a sort of "intrinsic strain" simply due to the different lattices of materials that have been piled up without directly stretching or compressing the layers.

In some examples, heterostrain (see Fig. 5.3) is also implemented in a different way exploiting the growth of two adjacent materials. In these cases, the mismatch

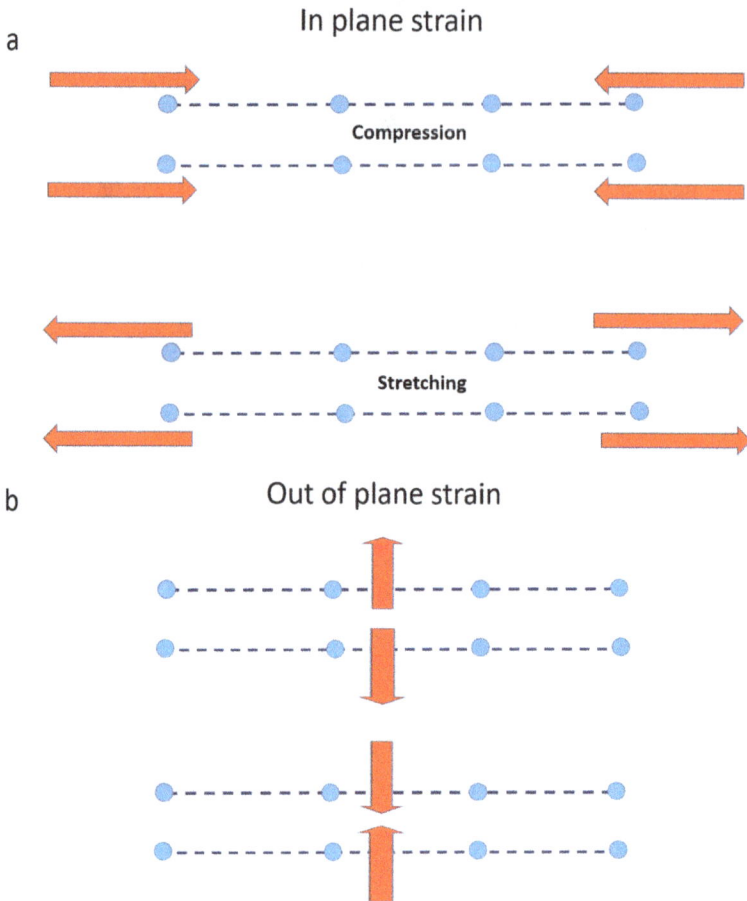

Fig. 5.2: (a) In-plane and (b) out-of-plane strain applied to a van der Waals structure composed of two same layers piled up.

between the two different materials is exploited to create a heterojunction with specific electronic properties as it will be shown in the next paragraphs (see Fig. 5.3).

5.3 In-plane strain: some examples in different fields

Concerning the in-plane strain, it has been exploited mainly to engineer the electronic band structure altering the covalent bond length and angles between the atoms in the 2D structure of the layers. One of the first pioneering works in the field was performed on graphene by Ni et al. in 2008 [6]. In this contribution, scientists demonstrated that applying a uniaxial stretching strain on graphene deposited on a flexible substrate "creates" a band in graphene that could be modulated as a function of the strain

Hetero-Strain

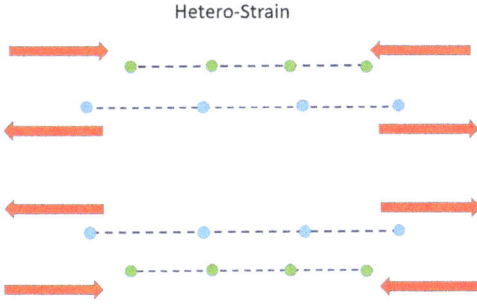

Fig. 5.3: Schematic view of heterostrain induced by piling up two layers of different origins. The different colours (green and blue) on the atoms highlight that we are using two different layers with different lattice constants (the distance between two of the atoms).

applied (Fig. 5.5). The change of the band gap value could be monitored simply using Raman spectroscopy as highlighted in Fig. 5.6. The change of gap is due to the elongation of the carbon bond in the strain direction. Some calculations predicted the opening of a gap of around 300 meV. Theoretically, the widening of the band gap, under the effect of uniaxial or shear strain, is due to the shift of the Dirac cones located at points K and K' [7, 8] in opposite directions. Therefore, if the uniaxial or shear strain is larger enough, the two inequivalent Dirac points, which move away from the K and K' points, respectively, may approach each other, and combine, culminating in the rising of a non-negligible band gap in graphene [9–11].

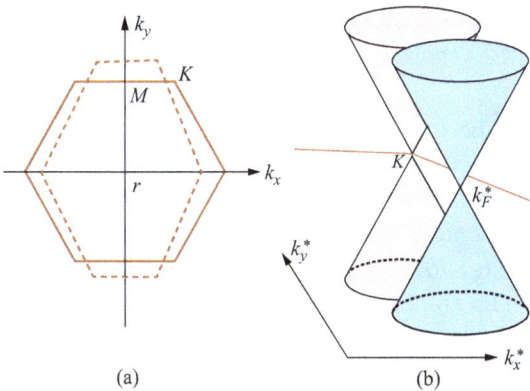

(a)   (b)

Fig. 5.4: Schematic illustrations of the effects of strain on graphene: (a) the deformation of the Brillouin zone; (b) the shifting of the Dirac cone away from the K point in the k^* space (reproduced with permission from [8], Copyright © 2010, Tsinghua University Press and Springer-Verlag, Berlin, Heidelberg).

This is a very interesting result, confirmed by other theoretical and experimental works [12, 13], considering that if this gap is stable at ambient temperature, we can imagine an interesting perspective to exploit graphene as a topological insulator at room temperature creating a gap through strain.

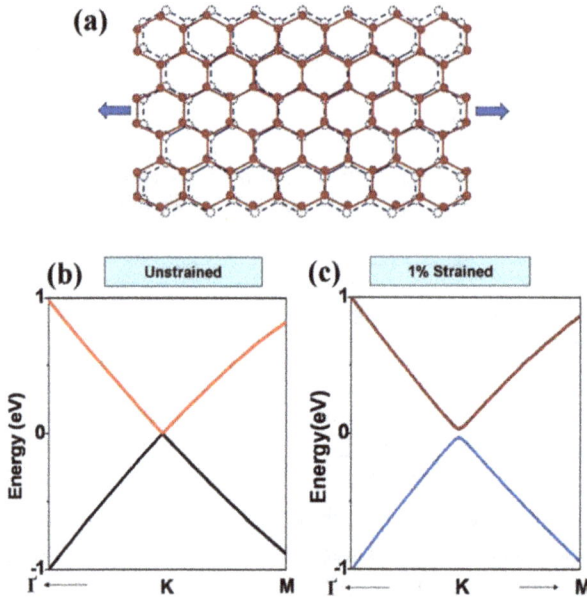

Fig. 5.5: (a) Schematic representation of the effect of uniaxial tensile stress on a graphene super-cell. (b) Schematization of strain effect before and after the strain is applied [16] (reprinted (adapted) with permission from [16], Copyright (2008) American Chemical Society).

As we have observed, graphene has been logically the first 2D material studied using straintronics. Then by subsequent emerging of other 2D materials, many teams decided to focus their attention on a larger panel of materials. We can mention, for example, the paper of Yang et al. in 2017 [15]. In this contribution, scientists developed a new test set-up that allows applying for the first time, biaxial strain, not like in the previous paper on graphene, in layers deposited on a flexible substrate exploiting a smart so-called blown-bubble technique. This technique consisted in applying a deformation on a PDMS layer used as a substrate by injecting gas under the substrate (see Fig. 5.7). Using this technique, they tested single- and two-layered MoS_2 structures and reported, thanks to Raman spectroscopy, the change/modulation of the band gap up to the failure of the layers.

We observe in Fig. 5.8(b) that the E^1_{2g} peak clearly splits into two peaks as soon as the strain is large enough because it modifies the lattice structure and breaks the inversion symmetry [16]. The Raman peak splitting points out that the strain showed an evident anisotropy reaching high value, which is mainly due to

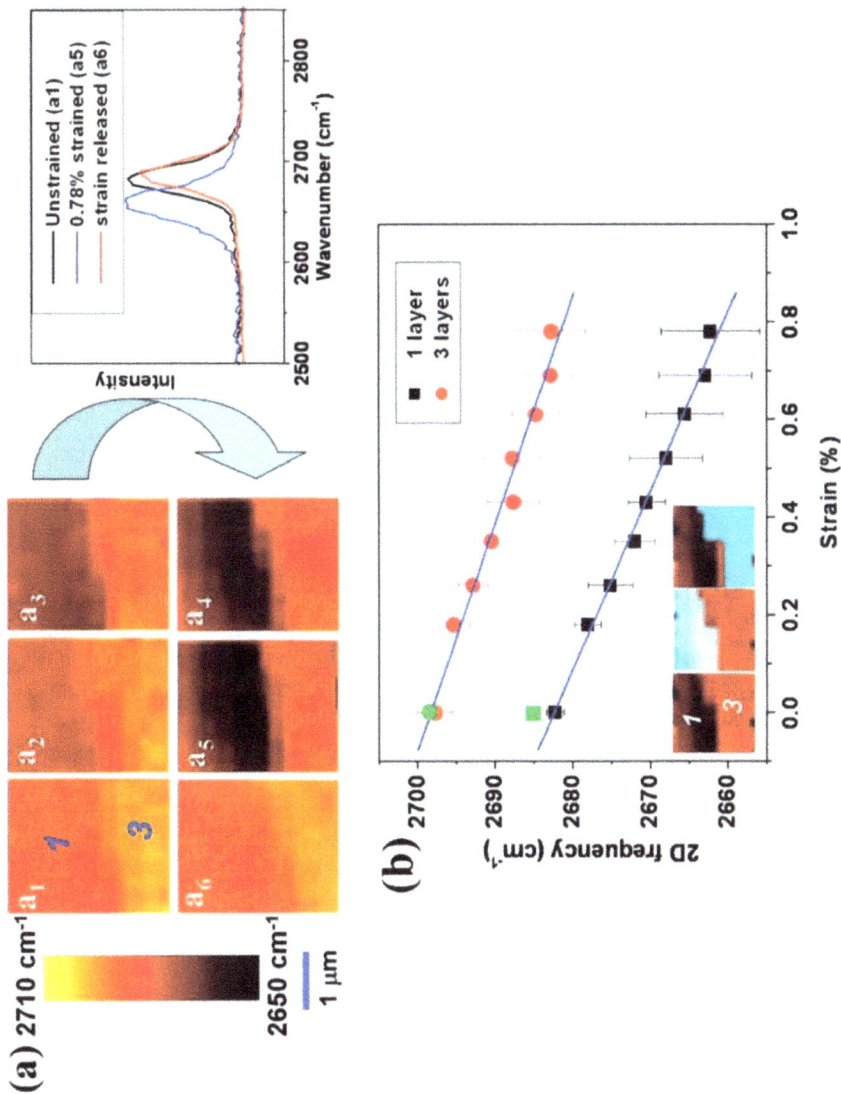

Fig. 5.6: (a) Two-dimensional frequency Raman images of unstrained and relaxed graphene samples. (b) The analysed 2D band frequency of single-layer (black squares) and three-layer (red circles) graphene under different uniaxial strains. The green square/circle is the frequencies of relaxed graphene [14] (reprinted (adapted) with permission from [16], Copyright (2008) American Chemical Society).

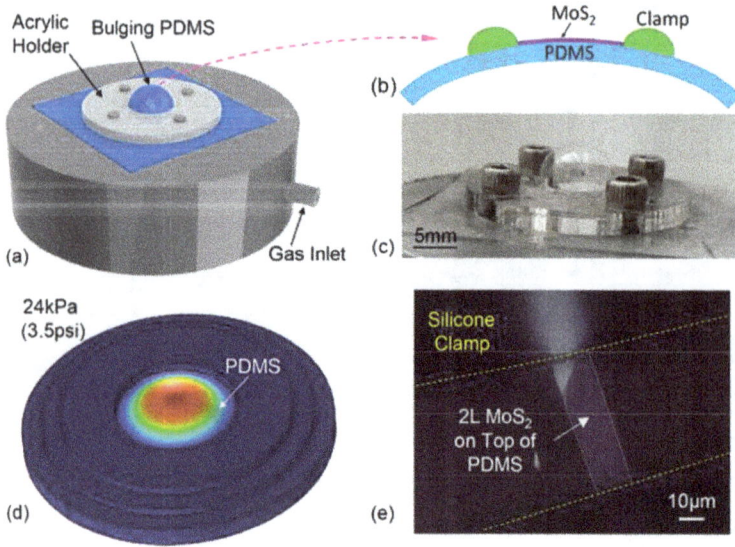

Fig. 5.7: Experimental set-up and representative MoS$_2$ device for blown-bubble bulge measurement. (a) Three-dimensional illustration of the experimental set-up. (b) Cross section of the MoS$_2$ device. (c) Photograph of the bulging PDMS observed during the experiment. (d) Finite element method simulation of the PDMS deformation under a differential gas pressure. (e) Optical image of a clamped 2L MoS$_2$ device (reprinted with permission from [15], Copyright (2017) American Chemical Society).

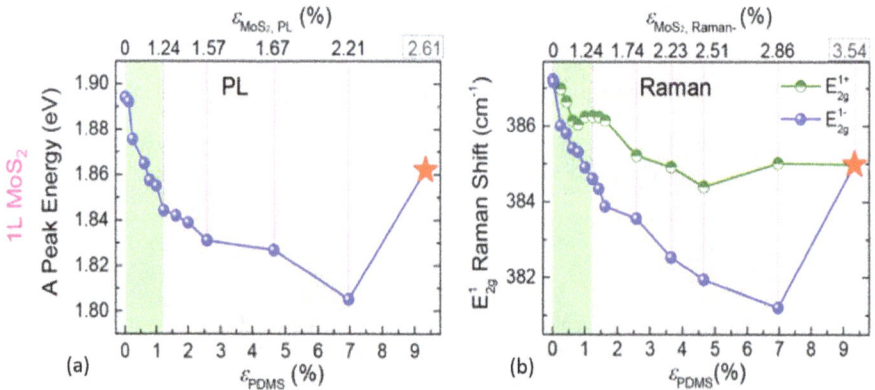

Fig. 5.8: Extracted (a) PL peak A and (b) Raman peak E^1_{2g} position shift with applied strain in the PDMS substrate, with the corresponding strain in MoS$_2$ indicated on the top axis. We can observe that the Raman spectroscopy highlights the sliding of the MoS$_2$ layer and so the anisotropy of the strain applied (reprinted with permission from [18], Copyright (2017) American Chemical Society).

the sliding between MoS$_2$ and the flexible substrate, specifically in the unclamped direction. This last observation puts in evidence that exploiting asymmetry in strain, structures with anisotropic characteristics can be achieved. This could be extremely useful in achieving multifunctional architecture of 2D materials by simply exploiting the strain response in different axes.

In a following paper by Zhang et al. in 2017 [17], scientists observed in black phosphorus that they could modulate in a continuous way through the application of uniaxial strain. Indeed, black phosphorus has a gap depending on the number of layers piled up (as shown in the previous chapter). This makes this material very interesting mainly for opto-electronic applications. One of the main peculiarities of this work was that scientists tested a device and not simply the response of a layer to strain (see Fig. 5.9). Indeed, they fabricated a transistor on a flexible substrate and observed the modifications of the electrical behaviour related to the applied strain.

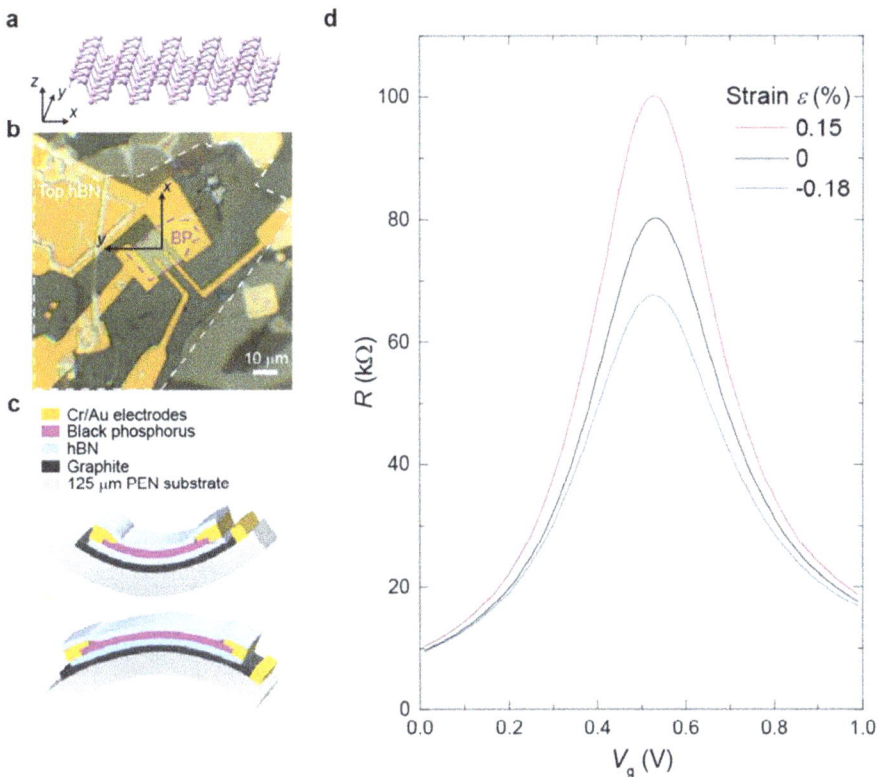

Fig. 5.9: Device configuration and strain-dependent ambipolar transport in black phosphorus (BP) FET. (a) Crystal structure of monolayer BP. (b) Optical image of a BP FET supported on a flexible substrate. (c) Schematic structure of a BP FET on flexible polyethylene naphthalate substrate under compressive/tensile strain. (d) Ambipolar transport behaviour of a BP FET under three different strain values (reprinted with permission from [17], Copyright (2017) American Chemical Society).

Magnetic effect can also be correlated to strain. Indeed, in the theoretical work of Zhang and co-workers in 2017 [18], they discovered that single-layer Fe_3GeTe_2 displays a strong uniaxial magnetocrystalline anisotropy energy of 920 µeV per Fe atom created by spin–orbit coupling (SOC). Scientists demonstrated that by applying biaxial tensile strains enhances the anisotropy energy and brings to light a strong magnetostriction in single-layer Fe_3GeTe_2 with a magneostrictive coefficient that cannot be neglected (see Fig. 5.10).

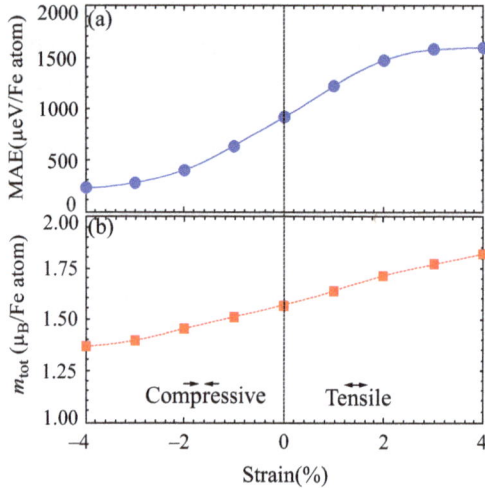

Fig. 5.10: Variation of (a) magnetic anisotropy energy (MAE) and (b) total magnetic moment per Fe atom of single-layer Fe_3GeTe_2 under biaxial strain calculated using the DFT-LDA approach levels (reprinted with permission from [18], Copyright (2016) American Physical Society).

These results pointed out that single-layer Fe_3GeTe_2 could be used in a useful way for magnetic storage applications. An essential parameter for magnetic recording materials (as shown in Fig. 5.10) is the magnetic anisotropy energy (MAE), which corresponds to the energy dependence as a function of the direction of magnetization. MAE is a concept used to describe some magnetic phenomena where the value of the magnetization vector does not change in a significant way while its direction varies notably. Indeed, in the framework of applications implying storing, it is easier to implement them having an easy magnetization axis and a consequent MAE. An extremely timely application of strain for vdW structures is in the field of topological matter [20–23]. Actually, in-plane strain could strongly influence the topology of the quantum states correlated to the rotational symmetry. This can achieve the topological switching in vdW topological materials. A highly interesting theoretical paper in this field has been published by Li et al. in 2015 [24]. Scientists employing ab initio electronic calculations were able to identify a new type of 2D topological insulator. They demonstrated the change from a trivial topological insulator phase to a topological insulator non-trivial one when monolayer of low-buckled HgTe and HgSe underwent an in-plane tensile strain of 2.6%, thanks to the coupling of strain and SOC effects.

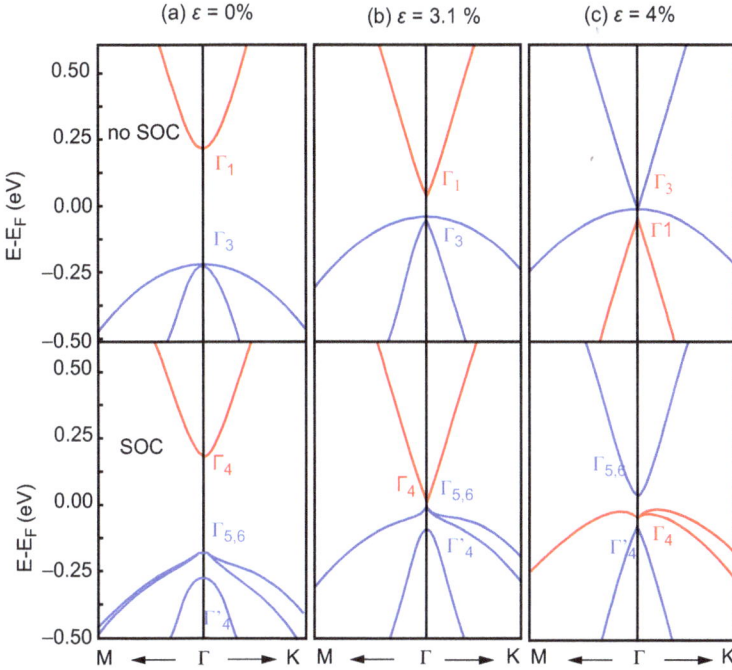

Fig. 5.11: Evolutions of the band structures of monolayer HgSe under in-plane tensile strain of (a) $\varepsilon = 0\%$, (b) 3.1% and (c) 4%. The band structures without and with SOC are shown in the upper and bottom panels, respectively (reprinted with permission from [24], Copyright (2015) Springer Nature).

In Fig. 5.11, it is shown how the uniaxial strain applied can lead to the inversion of the band and therefore to a non-trivial topological insulator phase for a tensile strain larger than 3%. Li and co-workers also discovered that the band gaps could be widened (0.2 eV for HgTe and 0.05 eV for HgSe) exploiting tensile strain, reaching results that largely exceeded those of experimentally obtained 2D quantum spin Hall insulators (see Chapter 1). These results were impressing considering that these new types of material could be suitable for practical applications of 2D topological insulators at room temperature, which is a booming and timely field of research. The main advantages of these materials are that teams need to have a great experience in growing them compared to the new generation of topological insulators such as stanene and plumbene. These last have been only very recently grown (in 2016 and in 2019, respectively, see Chapter 1) and are wholly artificial 2D materials. Again in the field of topological features, in a very recent similar work, Li et al. in

2021 [25] studied intrinsic 2D antiferromagnetic (AFM) insulators. They were able to show that a strain-engineered topological phase transition could be achieved in the 2D AFM topological insulator phase in $EuCd_2Sb_2$, thanks to in-plane magnetization. After first-principles calculations, the band gaps of $EuCd_2Sb_2$ quintuple layers could be modified, and a band gap closing/reopening process was also displayed for small critical tensile strain of 2%. The topologically non-trivial characteristics of strained $EuCd_2Sb_2$ QLs, with the opening of the gap, were strengthened with the help of direct calculation of the spin Chern number and topologically invariant $\mathbb{Z}_2$, and of the non-trivial topological nature of the edge states. One of the most interesting features of these new materials was that while the previously proposed magnetic topological states may be heavily deformed by fragile magnetism (see Chapter 1), the obtained 2D AFM topological insulator phase is highly robust against magnetic configurations, including ferromagnetic and AFM coupling with both in-plane and out-of-plane directions. This result is a major step to achieve and implement in real systems, devices exploiting robust topology phenomena, which has been up to now the main drawback that has slowed the progress of this research domain. Indeed, the influence of strain can be exploited as a sort of switch to reveal the topological insulator features in a quite deterministic way. Finally, the in-plane strain can be used to explore the superconductivity, for example, in case of magic angle structures with two or more graphene layers piled up. Indeed, we can imagine to achieve a superconducting anisotropic structure exploiting uniaxial strain in different directions. A very recent published work by Stephen and co-workers in 2021 [26] shed a light on the effect of uniaxial strain on the interacting phase diagram of magic angle twisted bilayer graphene and on the results obtained previously on these structures. Using both self-consistent Hartree–Fock and density matrix renormalization group calculations, scientists were able to highlight that small strain values are also applied ($\varepsilon \sim 0.1$–0.2%), which led to a zero-temperature phase transition between the symmetry-broken "Kramers intervalley coherent" insulator and a nematic semimetal [27]. The critical strain theoretically calculated lies in the interval of experimentally observed strain values; therefore, they predicted that strain was surely one of the reasons why the results were not completely reproducible and dependent on the measurement set-up with which the magic angle was obtained (the precision in obtaining a specific Moiré super-pattern in a completely deterministic way rotating two layers without any external influence such as the undesired strain). This observation can have a stronger impact on the trendy topic of magic angle and lead to an important questioning about the results recently obtained by a large panel of scientific teams working in this field. This questions also on the suitability of magic-angle-related structures considering that it is very difficult to deal in a completely reproducible way with strain in this nanometric structures. Moving to superconductivity, Lin et al. published in 2015 [28] a very interesting contribution where they explored the quasi-2D superconductivity in $FeSe_{0.3}Te_{0.7}$ thin films and their tuning exploiting tensile strain (see Fig. 5.12). In fact, they observed the structural and superconducting properties of $FeSe_{0.3}Te_{0.7}$ (FST) thin films with different thicknesses grown on ferroelectric $Pb(Mg_{1/3}Nb_{2/3})_{0.7}Ti_{0.3}O_3$ substrates. Knowing that the FST films underwent biaxial tensile strains which were fully relaxed for films thicker than 200 nm.

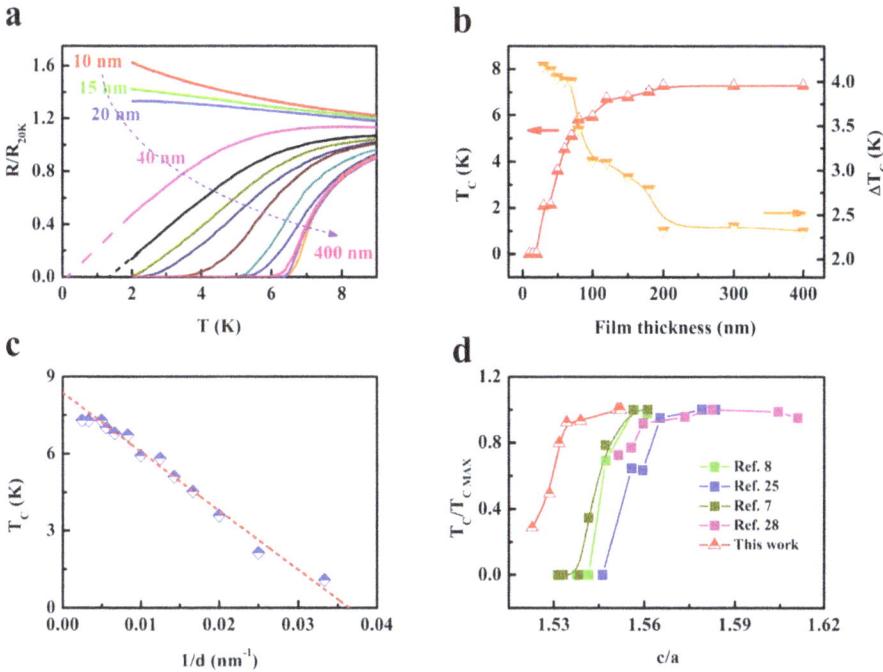

Fig. 5.12: Electrical transport properties of FST films with different thicknesses. (a) T curves at low temperatures for FST films with different thicknesses. (b) Variation of the critical temperature and its change with film thickness. (c) Plot of the critical temperature as a function of $1/d$. (d) Variation of critical temperature for FST films with c/a (reprinted with permission from [28], Copyright (2015) Springer Nature).

Electrical transport measurements unveiled that the ultrathin films exhibited an insulating behaviour and superconductivity appeared for thicker films with a critical temperature (the temperature under which a material acts as a superconductor), saturated above 200 nm. Under the effect of an electric field applied to a heterostructure, the critical temperature of FST thin film increased because of the tensile strain reduction. As highlighted, also in this case, it has been demonstrated that the strain can be used as a sort of trigger to switch on or off specific properties of the 2D (also thin films) materials.

5.4 Out-of-plane strain

The interlayer coupling that characterized vdW structure is relatively weak compared to intra-layer binding due to its, by definition, vdW nature. Indeed, in the
second case we are dealing with largely stronger chemical bond. Actually, many
physical properties sensitive to the interlayer spacing have logically been engi-
neered via the out-of-plane strain considering that it is perpendicular to the stacked
layers. In one of the first work on graphene, it has been outlined that the out-of-
plane strain can introduce positive piezoconductive effect in few layered graphene,
which is inaccessible through the in-plane strain engineering. Xu and co-workers in
2014 [29] set up a bench to perform in situ piezoconductive measurements, exploiting a pressure-modulated conductance microscopy [30, 31], with a non-conducting
atomic force microscopy tip (see Fig. 5.13). Thanks to that, scientists obtained a to-
pography image of the graphene layers undergoing the effect of the tip pressure inducing out-of-plane strain that could be modulated. In order to be more precise, the
strain applied using the atomic force microscopy probe is perpendicular to the
layers; however, it also induces bi-axial strain in the direction parallel to the layers
because the layers are clamped in the extremities and also because the tip can act
only locally deforming the graphene membrane in 2D. For this reason, as already
explained, it is necessary to evaluate with attention the total strain and its direction
globally in order to achieve a suitable results analysis.

Scientists highlighted that this piezoconductive effect was strictly related to the
number of graphene layers piled up and that the most pronounced response was for
a trilayer graphene sample. The effect, and its dependence on the layer number,
was explained resulting from the strain-induced competition between interlayer
coupling and intralayer transport. Moving to another field of research and more specifically on "magic angle" (see Chapter 2), a very interesting work by Yankowitz
et al. in 2019 [32], again on graphene vdW structures, demonstrated that the value of
the magic angle could be tuned to a value larger than 1.1°, inducing superconductivity at larger angles, and this simply applies a hydrostatic pressure perpendicularly
in a way that the interlayer coupling can be varied to fine-tune these phases. This
phenomenon could be associated logically with the reduction of interlayer spacing
induced by out-of-plane strain in magic angle twisted bilayer graphene and therefore reducing the interlayer hopping energy, leading to the shift of the magic angle
for the superconducting and correlated insulating states. From a long-term point of
view, we can imagine to use the out-of-plane strain to induce a robust topological
insulator phase in bilayer structures composed of graphene and transition metal dichalcogenides (TMDs). A study by Island et al. in 2019 [33] demonstrated that,
thanks to the proximity effect, this two-layer structure led to an incompressible
gapped structure at charge neutrality. Considering that SOC is the main parameter to
achieve the time-reversal invariant topological phases of matter, the experimental
data agreed quantitatively with a simple theoretical model in which the new phase

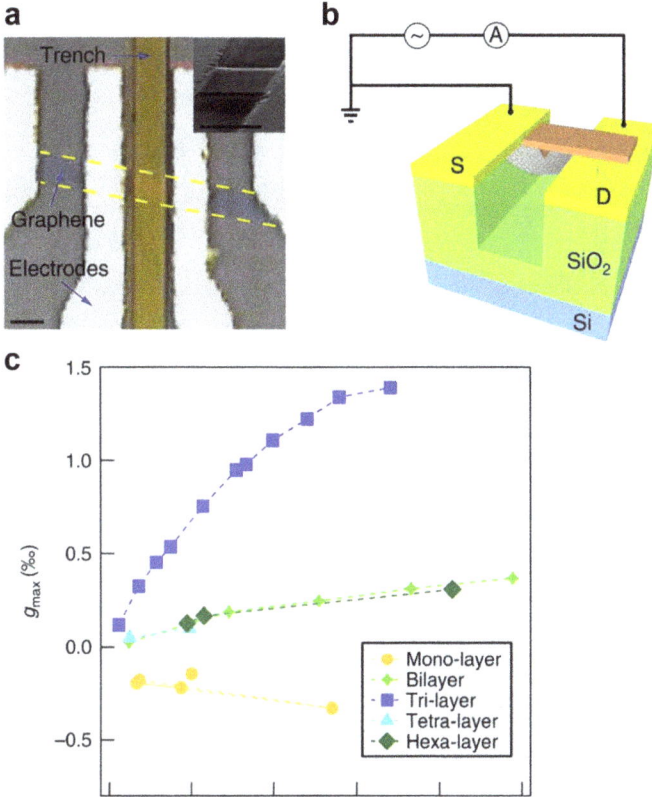

Fig. 5.13: Suspended graphene device and PCM set-up. (a) Optical microscopic image of a four-terminal suspended bilayer graphene device. (b) Schematic set-up of pressure-modulated conductance microscopy (PCM) performing piezoconductive measurements on suspended graphene. (c) Plot of the maximum relative conductance change g_{max} as a function of strain for various suspended graphene devices (reprinted with permission from [29], Copyright (2015) Springer Nature).

was the typical result of a SOC-driven band inversion, typical for topological insulators (see Fig. 5.14). Curiously, and in contrast to Kane–Mele SOC in monolayer graphene (see Chapter 1), the inverted phase was not confirmed to be a time-reversal-invariant topological insulator, despite being separated from conventional band insulators by electric-field-tuned phase transitions where crystal symmetry implied that the bulk gap had to vanish [34].

The electrical transport measurements performed by Island and co-workers pointed out that the inverted phase had a conductivity of approximately $e2/h$, which vanished when it endured extremely small in-plane magnetic fields. The high conductivity and anomalous magnetoresistance were found consistent with theoretical models that predicted the existence of helical edge states within the inverted phase

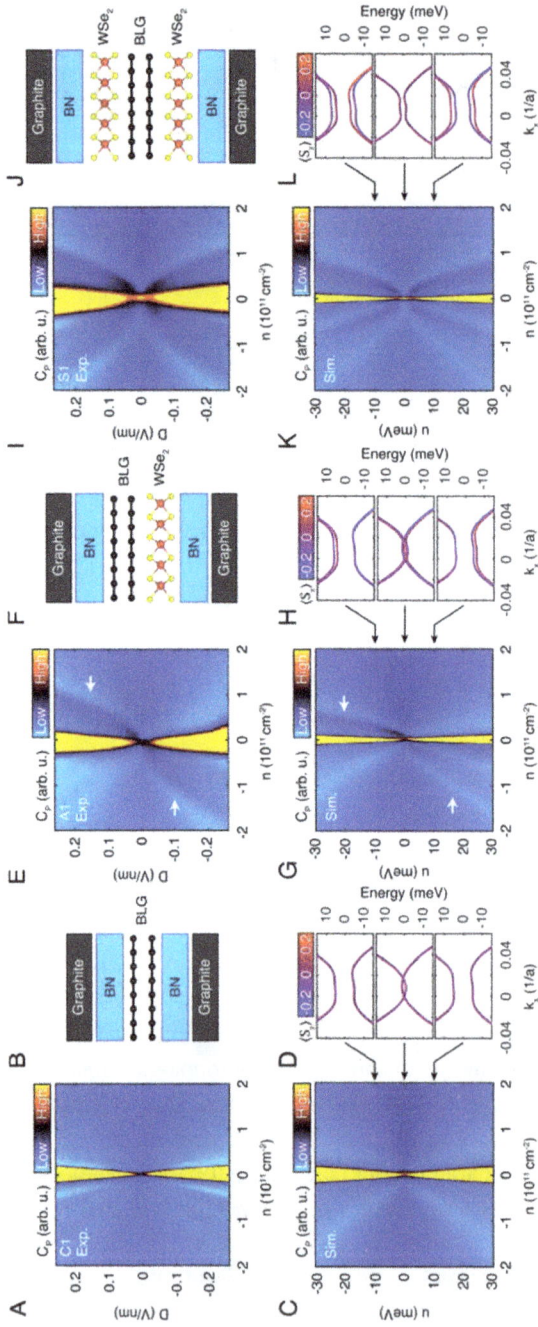

Fig. 5.14: Comparison between experimental and simulated penetration field capacitances for the three device configurations studied. The structures studied are shown at the left of each graph (B, F, J). A scheme showing the band inversion as an effect of the layers proximity is shown in (D, H, L). For more details check the reference (reprinted with permission from [33], Copyright (2019) Springer Nature).

protected from backscattering by an emergent spin symmetry that stayed robust even for large Rashba SOC. These results are extremely promising and pave the way for using layer proximity engineering, also exploiting out-of-plane strain, of robust topological insulators as well as correlated quantum phases in the strong spin–orbit regime in graphene heterostructures. The final dream is that this new methodology could lead to the achievement of vdW structures, enabling robust topological insulator features at room temperature to push their implementation in real-life devices. Again, exploiting the layer proximity, Zheng et al., members of Jarrillo-Herrero's team at MIT in 2020 [35], observed an emergent ferroelectricity in graphene-based Moiré heterostructures (see Fig. 5.15). In fact, it is not apparently logic to expect that graphene could show ferroelectricity considering its intrinsic homogeneous composition (same atoms in hexagonal networks).

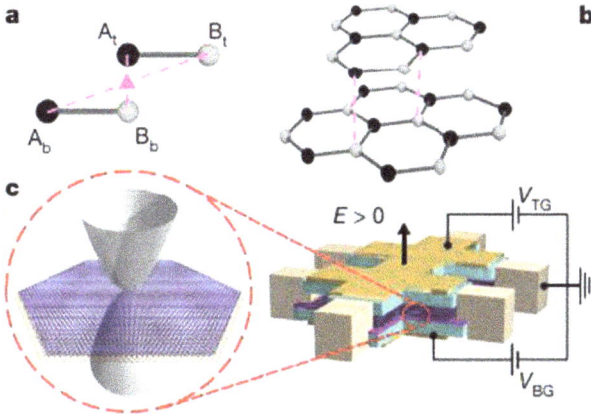

Fig. 5.15: (a) Lattice structure of Bernal-stacked bilayer graphene. (b) Schema of the superposition of the two-layer lattices. (c) Schematic of BN-encapsulated bilayer device. The black arrow identifies the electric field direction (reprinted with permission from [35], Copyright (2020) Springer Nature).

However, scientists were able to achieve switchable ferroelectricity in Bernal-stacked bilayer graphene lying between two hBN layers. This effect seemed to be related to the introduction of a Moiré superlattice [36, 37] potential (via aligning bilayer graphene with the top and/or bottom boron nitride crystals), which allowed the appearance of a robust hysteretic behaviour of graphene resistance because of an externally applied out-of-plane displacement field. These results led to the disclosure of an unconventional, odd-parity electronic ordering in the bilayer graphene/boron nitride Moiré system. The observed Moiré ferroelectricity could enable ultrafast, programmable and atomically thin carbon-based memory devices simply exploiting two layers of graphene. This phenomenon can be enhanced applying the out-of-plane strain and could lead to a deterministic control of the effect.

5.5 Heterostrain

As highlighted by its name, heterostrain is a form of strain that is created in materials, and specifically 2D materials, which have a different composition and lattice and are piled up or grown on the same substrate [4]. Heterostrain is mainly obtained through lattice mismatch and is inhomogeneous. It is the largest at the centre of the mismatched interface, while it decreases in the region far away from the interface. To inhibit it, usually the key parameters are different Young's modulus or the thermal expansion coefficient of the material. The lattice deformation can be spotted using high-resolution microscopic characterization methods, such as scanning transmission electron microscope and scanning tunnelling microscopy (STM). Thanks to these techniques, a detailed mapping of the surface devices can be performed. This kind of strain is ubiquitous in vdW heterostructures. As already outlined, the heterostrain has the capability of tuning intralayer hopping/exchange energy and breaking rotational symmetry. However, heterostrain owns many features similar to the interlayer strain too. Actually, heterostrain could modulate the interlayer coupling by altering the relative displacement of the adjacent layers. Thanks to a potential heterostrain engineering (also called in some case Moiré engineering), potentially we can think to achieve a platform to explore electronic phenomena and photonic or topological properties. Heterostrain, usually in vdW heterostructures, is due to the inevitable thermal relaxation during the layer-stacking process. This process leads to inhomogeneous and uncontrollable strain distribution. Such a disorder of heterostrain distribution is not only an important obstacle to the deterministic experimental investigation on the role of heterostrain engineering on correlated electronic behaviours in twisted bilayer graphene and TMDs, but also strongly reduces the extension of heterostrain engineering to other types of vdW heterostructures. Indeed, considering that heterostrain is not verifiable, even if it is very low it can influence in a dramatic way the interpretation of experimental data. This is particularly true for measurements of the magic angle where the reproducibility of the fabrication of structures is surely the main drawback that limits a potential implementation of the phenomenon in real devices. To that end, developing new techniques to fabricate uniform and designable heterostrain in vdW heterostructures is one of the major challenges. A different way to implement heterostrain is to use the mismatch in creating heterojunction between two different 2D materials. Examples of this technique were presented by the works of Zhang et al. in 2018 [38] and of Li et al. in 2015 [39], both are of the same scientific team. In these pioneering works, scientists reported the two-step epitaxial growth of lateral WSe_2–MoS_2 heterojunction, where the edge of WSe_2 induces the epitaxial MoS_2 growth despite a large lattice mismatch. A work with some analogies has been performed by Zhang and co-workers [40] (see Fig. 5.16). Scientists explored the heterojunction using STM and identified the Moiré pattern to analyse the effects of the full 2D strain tensor with high spatial resolution. They were able to observe the band alignment of the WSe_2–MoS_2 lateral heterojunction and that the misfit strain is

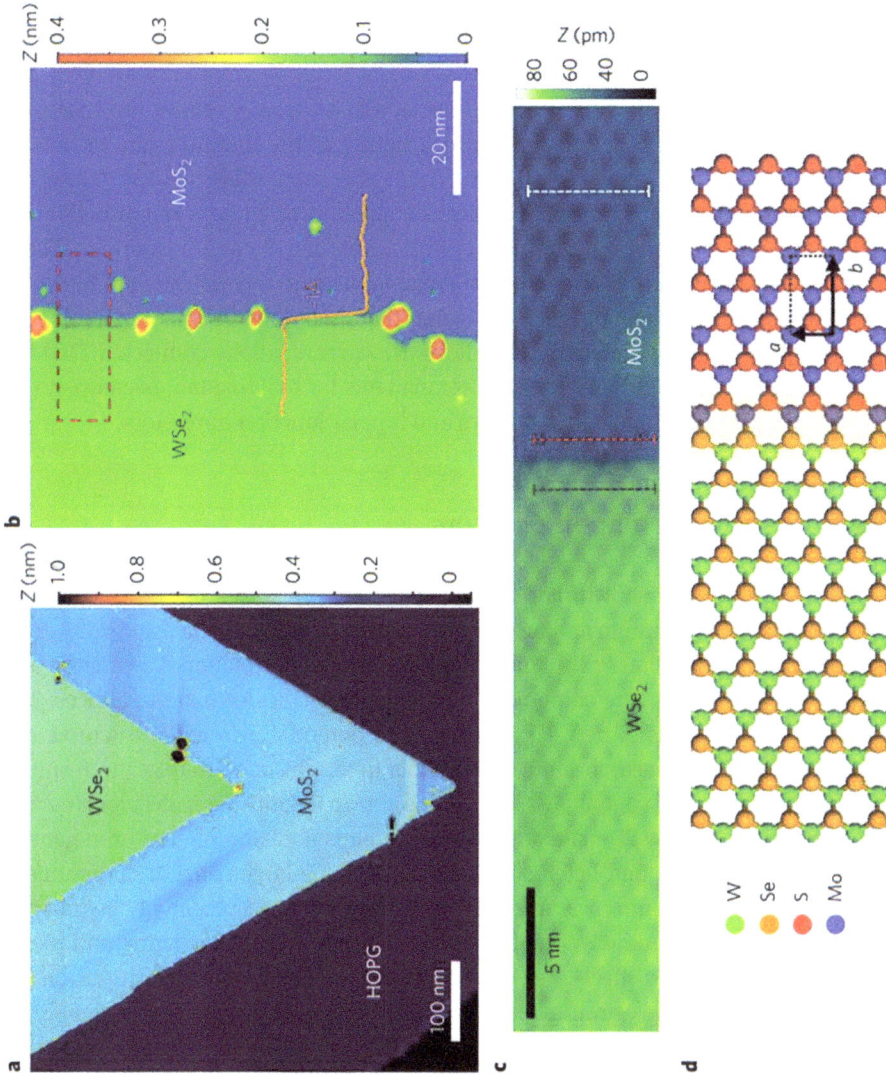

Fig. 5.16: STM images of the WSe$_2$–MoS$_2$ lateral heterojunction. (a) Large-scale STM image showing that the inner WSe$_2$ core is surrounded by an MoS$_2$ skirt. (b) Close-up image of the interface and the thickness difference between WSe$_2$ and MoS$_2$. (c) Zoomed-in image of a straight interface segment. (d) Model of the WSe$_2$–MoS$_2$ heterojunction with a W–S line interface (reprinted with permission from [40], Copyright © 2018, Nature Springer Publishing Group).

the main cause of type II to type I band alignment change. Indeed, this was performed, thanks to the lattice distortion observation that allowed determining the full strain tensor.

STM reveals the dislocations at the interface that partially relieve the strain. Finally, we observe a distinctive electronic structure at the interface due to hetero-bonding.

The epitaxial growth process has been demonstrated to be a reproducible approach to achieve lateral heterojunction with an atomically sharp interface. These structures did not own the advantages of being fabricated only piling up different 2D layers, and also it is commonly accepted that the direct growth is more suitable for the thermodynamically preferred TMD alloys. However, spatially connected TMD lateral heterojunctions could be important components for building monolayer p−n rectifying diodes, LEDs, photovoltaic devices and bipolar junction transistors.

5.6 Conclusions and perspectives

Summarizing, we have shown that strain engineering provides a powerful means for tailoring electronic, quantum and topological properties in 2D materials and vdW structures by directly modulating the interatomic distances and rotational symmetry and more generally through stretching or compressing molecules/compound lattices. The discovery of thousands of new 2D materials and potential new vdW structures has largely widened the research landscape in this field. This also thanks the result of modelling through new computational methods able to identify the molecular structure and composition of new 2D materials (see, e.g. what is done for MXenes or in general for new materials [41]), and heterostructures disclosing a large panel of opportunities for the research of straintronics. All these materials that could potentially be immediately investigated include vdW semiconductors for band gap engineering and piezo-conductive/piezoelectric effects, heterostructures consisting of vdW semiconductors for specific excitonics/optoelectronics, optimized twisted bilayer graphene or strain Moiré-engineered TMD structures showing superconductivity at higher temperatures, vdW composed of stacking magnetic materials where strain can perform switching between different behaviours and, finally, maybe the most important considering the impact of this topic in the future, switching of topological features of vdW or more generally 2D materials. We can mention the possibility of exploiting uniaxial strain applied to graphene to create the basis for higher temperature demonstration of topological insulator features that cannot be proved with intrinsic graphene considering its too low SOC. This could be a major breakthrough for science. We can conclude that, straintronics, as an important strategy for finely tuning the electronic structure and properties of 2D materials, has recently focused the research interests of many research teams. Indeed, there are a variety of methods to induce strain to 2D materials due to their atomic-scale thicknesses and strong in-plane and out-of-plane deformation capacity. Moreover, as

a matter of fact, 2D materials are more sensitive to strain compared to the bulk structure. As a consequence, small strains can significantly change the 2D's lattice structure,
modulating their physical properties. Thanks to that, straintronics could enhance in
an extensively field of application of 2D materials for flexible strain sensors, flexible photodetectors and other wearable optoelectronic devices but also for and
more specifically in a beyond CMOS perspective [42], where the architecture of
new electronic components based on topology features of 2D/vdW materials can
be triggered by opportune strain engineering. However, a lot of technical hurdles
need to be overcome. It is difficult at nanometric level to achieve extremely reproducible strains. Indeed, the interfacial interaction strength between the substrate and
2D materials is not completely understood, resulting that the strain in the deformed
substrate cannot be effectively transferred to 2D materials. As a result, it is particularly important to study the interface mechanical behaviour between 2D materials
and the substrate, especially the interface adhesion and friction behaviours. The
methods to induce strain in 2D and vdW structure have to be improved and standard-
ized to think to wholly exploit strain engineering. If we will not able to do that this
technology for 2D will remain a curiosity for lab tests and will not impact markets.
Another drawback is a lack of a deeper understanding of the strain-related phenomena. It is necessary to further understand the intrinsic relationship between structure
and property of 2D materials under the action of strain and to completely understand
how to modulate in a deterministic way the 2D/vdW properties. It is also evident that
the integration of strain, and how to modulate it, in real devices is another technical
hurdle that has to be tackled and has not been sufficiently studied.

References

[1]   https://goldbook.iupac.org/terms/view/V06597
[2]   https://www.manchester.ac.uk/discover/news/playing-lego-on-an-atomic-scale/
[3]   Thompson, S. E., Armstrong, M., Auth, C., et al. A 90-nm logic technology featuring strained-
       silicon. IEEE Trans Electron Dev. 51, 11, 2004, 1790–1797.
[4]   Miao, F., Liang, S. J., Cheng, B. Straintronics with van der Waals materials. Npj Quantum
       Mater. 6, 59, 2021, https://doi.org/10.1038/s41535-021-00360-3.
[5]   Strain engineering of two-dimensional materials: Methods, properties, and applications,
       Yang, S., Chen, Y., Jiang, C. InfoMat, 3, 4, 2021, pp. 397–420, https://doi.org/10.1002/inf2.
       12177
[6]   Ni, Z. H., Yu, T., Lu, Y. H., Wang, Y. Y., Feng, Y. P., Shen, Z. X. Uniaxial Strain on Graphene:
       Raman Spectroscopy Study and Band-Gap Opening. ACS Nano. 2, 11, 2008, 2301–2305.
[7]   Cocco, G., Cadelano, E., Colombo, L. Gap opening in graphene by shear strain. Phys. Rev.
       B. 81, 2010, 241412. https://doi.org/10.1103/PhysRevB.81.241412.
[8]   Li, Y., Jiang, X., Liu, Z., et al. Strain effects in graphene and graphene nanoribbons: The
       underlying mechanism. Nano Res. 3, 2010, 545–556, https://doi.org/10.1007/s12274-010-
       0015-7.

[9] Choi, S.-M., Jhi, S.-H., Son, Y.-W. Phys. Rev. B. 81, 2010, 081407. https://doi.org/10.1103/PhysRevB.81.081407.

[10] Farjam, M., Rafii-Tabar, H. Phys. Rev. B. 80, 2009, 167401. https://doi.org/10.1103/Phys RevB.80.167401.

[11] Sagalianov, I. Y., Radchenko, T. M., Prylutskyy, Y. I., et al. Mutual influence of uniaxial tensile strain and point defect pattern on electronic states in graphene. Eur. Phys. J. B. 90, 2017, 112, https://doi.org/10.1140/epjb/e2017-80091-x.

[12] Ohmi, Y., Ogawa, M., Souma, S. Effect of uniaxial strain on the electronic transport in single layer graphene, 2011 International Meeting for Future of Electron Devices, 2011, pp. 126–127, doi: 10.1109/IMFEDK.2011.5944877.

[13] Si, C., Suna, Z., Liu, F. Strain engineering of graphene. Nanoscale. 8, 2016, 3207–3217, https://doi.org/10.1039/C5NR07755A.

[14] Ni, Z. H., Yu, T., Lu, Y. H., Wang, Y. Y., Feng, Y. P., Shen, Z. X. Uniaxial strain on graphene: Raman spectroscopy study and band-gap opening. ACS Nano. 2, 11, 2008, 2301–2305, 2008 https://doi.org/10.1021/nn800459e.

[15] Yang, R., Lee, J., Ghosh, S., Tang, H., Sankaran, R. M., Zorman, C. A., Philip, X.-L. F. Tuning optical signatures of single- and few-layer MoS2 by blown-bubble bulge straining up to fracture. Nano Lett. 17, 8, 2017, 4568–4575. 2017 https://doi.org/10.1021/acs.nanolett.7b00730.

[16] Conley, H. J., Wang, B., Ziegler, J. I., Haglund, R. F., Pantelides, S. T., Bolotin, K. I. Nano Lett. 13, 2013, 3626–3630.

[17] Zhang, Z., Li, L., Horng, J., Wang, N. Z., Yang, F., Yu, Y., Zhang, Y., Chen, G., Watanabe, K., Taniguchi, T., Chen, X. H., Wang, F., Zhang, Y. Strain-modulated bandgap and piezo-resistive effect in black phosphorus field-effect transistors. Nano Lett. 17, 10, 2017, 6097–6103. 10.1021/acs.nanolett.7b02624.

[18] Zhang, Z., Likai, L., Horng, J., Wang, N., Yang, F., Yijun, Y., Zhang, Y., Chen, G., Watanabe, K., Taniguchi, T., Chen, X., Wang, F., Zhang, Y. Strain-modulated bandgap and piezo-resistive effect in black phosphorus field-effect transistors. Nano Lett. 17, 10, 2017, 6097–6103. 2017 10.1021/acs.nanolett.7b02624 2017.

[19] Houlong, L., Zhuang, P. R. C. K., Hennig, R. G. Strong anisotropy and magnetostriction in the two-dimensional Stoner ferromagnet Fe3GeTe2. Phys. Rev. B. 93, 134407.

[20] Topology on top. Nature Phys. 12, 2016, 615, https://doi.org/10.1038/nphys3827.

[21] Asorey, M. Space, matter and topology. Nat. Phys. 12, 2016, 616–618, https://doi.org/10.1038/nphys3800.

[22] Huber, S. Topological mechanics. Nature Phys. 12, 2016, 621–623, https://doi.org/10.1038/nphys3801.

[23] Topological Matter, Lectures from the Topological Matter School 2017, Editors Bercioux, D., Cayssol, J., Vergniory, M. G., Calvo, M. R. XVI, 261, Springer Series in Solid-State Sciences, https://doi.org/10.1007/978-3-319-76388-0

[24] Li, J., He, C., Meng, L. et al. Two-dimensional topological insulators with tunable band gaps: Single-layer HgTe and HgSe. Sci Rep. 5, 2015, 14115, https://doi.org/10.1038/srep14115.

[25] Li, R., Wang, H., Mao, N., Ma, H., Huang, B., Daia, Y., Niua, C. Engineering antiferromagnetic topological insulator by strain in two-dimensional rare-earth pnictide EuCd2Sb2. Appl. Phys. Lett. 119, 173105, 2021, https://doi.org/10.1063/5.0063353.

[26] Parker, D. E., Soejima, T., Hauschild, J., Zaletel, M. P., Bultinck, N. Strain-induced quantum phase transitions in magic-angle graphene. Phys. Rev. Lett. 127, 2021, 027601, https://doi.org/10.1103/PhysRevLett.127.027601.

[27] Liu, S., Khalaf, E., Lee, J. Y., Vishwanath, A. Nematic topological semimetal and insulator in magic-angle bilayer graphene at charge neutrality. Phys. Rev. Res. 3, 1, 12, 2021, 013033. https://link.aps.org/doi/10.1103/PhysRevResearch.3.013033.

[28] Lin, Z., Mei, C., Wei, L., et al. Quasi-two-dimensional superconductivity in $FeSe_{0.3}Te_{0.7}$ thin films and electric-field modulation of superconducting transition. Sci. Rep. 5, 14133, 2015, https://doi.org/10.1038/srep14133.

[29] Xu, K., Wang, K., Zhao, W., et al. The positive piezoconductive effect in graphene. Nat. Commun. 6, 8119, 2015, https://doi.org/10.1038/ncomms9119.

[30] Lau, C. N., Stewart, D. R., Williams, R. S., Bockrath, M. Direct observation of nanoscale switching centers in metal/molecule/metal structures. Nano Lett. 4, 2004, 569–572, 10.1021/NL035117A.

[31] Miao, F., Ohlberg, D., Stewart, D. R., Williams, R. S., Lau, C. N. Quantum conductance oscillations in metal/molecule/metal switches at room temperature. Phys. Rev. Lett. 101, 2008, 016802, https://doi.org/10.1103/PhysRevLett.101.016802.

[32] Yankowitz, M., Chen, S., Polshyn, H., Zhang, Y., Watanabe, K., Taniguchi, T., Graf, D., Young, A. F., Dean, C. R. Tuning superconductivity in twisted bilayer graphene. Science. 8, 363, 6431, 2019, 1059–1064, 10.1126/science.aav1910.

[33] Island, J. O., Cui, X., Lewandowski, C., et al. Spin–orbit-driven band inversion in bilayer graphene by the van der Waals proximity effect. Nature. 571, 2019, 85–89, https://doi.org/10.1038/s41586-019-1304-2.

[34] Zaletel, M. P., Khoo, J. K. The gate-tunable strong and fragile topology of multilayer-graphene on a transition metal dichalcogenide. Preprint at https://arXiv.org/abs/1901.01294 (2019).

[35] Zheng, Z., Ma, Q., Bi, Z., et al. Unconventional ferroelectricity in moiré heterostructures. Nature. 588, 2020, 71–76, https://doi.org/10.1038/s41586-020-2970-9.

[36] Cao, Y., Fatemi, V., Fang, S., et al. Unconventional superconductivity in magic-angle graphene superlattices. Nature. 556, 2018, 43–50, https://doi.org/10.1038/nature26160.

[37] Sun, Z., Hu, Y. H. How magical is magic-angle graphene?. Matter. 2, 5, 2020. Pages 1106–1114. https://doi.org/10.1016/j.matt.2020.03.010.

[38] Zhang, C., Li, M. Y., Tersoff, J., et al. Strain distributions and their influence on electronic structures of WSe_2–MoS_2 laterally strained heterojunctions. Nat. Nanotech. 13, 2018, 152–158, https://doi.org/10.1038/s41565-017-0022-x.

[39] Li, M. Y., Shi, Y., Cheng, C. C., Lu, L. S., Lin, Y. C., Tang, H. L., Tsai, M. L., Chu, C. W., Wei, K. H., He, J. H., Chang, W. H., Suenaga, K., Li, L. J. Nanoelectronics. Epitaxial growth of a monolayer WSe2-MoS2 lateral p-n junction with an atomically sharp interface. Science. 349, 6247, 2015, 524–528. 10.1126/science.aab4097.

[40] Zhang, C., Li, M. Y., Tersoff, J., et al. Strain distributions and their influence on electronic structures of WSe2–MoS2 laterally strained heterojunctions. Nat. Nanotech. 13, 2018, 152–158, https://doi.org/10.1038/s41565-017-0022-x.

[41] Vasudevan, R. K., Choudhary, K., Mehta, A., et al. Materials science in the artificial intelligence age: High-throughput library generation, machine learning, and a pathway from correlations to the underpinning physics. MRS Commun. 9, 2019, 821–838, https://doi.org/10.1557/mrc.2019.95.

[42] http://www.itrs2.net/

6 Exotic properties of 2D materials: where are we? The compromise between beyond CMOS and more-than-Moore vision and roadmaps

6.1 Premises to conclusion and to overall analysis

In this book, we have presented some of the most interesting new features recently highlighted in 2D materials and some strategies to make them emerge. Our approach has been, all along the book, to try to make understand the physics behind each phenomenon in an understandable way, using, as possible, simple terms. Indeed, in the papers recently published, it is quite difficult to have a clear global view of the physics behind exotic properties of 2D materials without developing difficult mathematical passages. Therefore, it is not easy to grab some hints about physics or about the potential applications in a quite rapid way. Moreover, it is impossible to give an exhaustive overview of all the phenomena presented. We would have needed to write a book for each one. But this was not our aim. Our objective was to give a rapid overview pointing out the main features. In the next sections, we will formulate some considerations about the different topics developed in the book and will present our vision. We will try to explicit some roadmaps that will be an easy tool to understand if some of the different fields will lead to major innovations. We do not have the arrogance to think that we know exactly what will happen and that we have an exhaustive knowledge of all the phenomena. We are dealing with fundamental physics, which is, for some of the topics, completely new. We will simply try to give some suggestions and to define which could be the potential implementation or future directions of the research. We hope we have been able to succeed in and we hope the reader enjoyed it.

6.2 Short introduction on the meaning of "Moore's law", "more-than-Moore" and "beyond CMOS" vision

Each technology can be classified as a function of its coherence with the Moore's law. But what is indeed the Moore's law so often quoted? Gordon Moore was the co-founder of Fairchild Semiconductor and Intel, whose 1965 seminal paper described a doubling every year in the number of components per integrated circuit and projected this rate of growth would continue for at least another decade [1, 2]. Up to now, this tendency has been accepted as an indisputable reality by semiconductor industry. This vision was nourished by the fact that producing chips bigger and transistors smaller meant that semiconductor companies have for decades invested on R&D, and the facilities have become much more expensive. However, considering the chips made out of

https://doi.org/10.1515/9783110656336-007

smaller transistors became faster and more capable, the market for them continued to grow allowing the chip manufacturers to recover their R&D costs increasing the efforts to make their products still tinier.

Fig. 6.1: Virtuous circle of chip manufacturers [3].

This was a very virtuous circle (see Fig. 6.1) that the companies assumed to be nearly never-ending. Now, things have changed. The limits of miniaturization will be reached in the near future (around 20 years) and it will not be possible to continue developing the same technologies based mainly on charge transport and modulation to achieve better performances of devices. The industry reaction was to issue the so-called more-than-Moore vision [4], where no breakthroughs in terms of concepts or fabrication are targeted but where new functionalities will be added to the existing devices or systems. It is a sort of parallel development of the Moore law. This approach is very interesting for industry that does not have to change their production lines but only to update them. Moreover, to produce devices with added functionalities is a quite smart approach to reach a large market. Mobile phone is a quite impressing example of this tendency considering that we are talking about a system where the main innovation is achieved by adding new functionalities simply optimizing existing technologies. Thanks to that companies can increase their gain with the lowest efforts in terms of technology adaptability. However, considering the great challenges of the next future, where more sustainable technologies have to be developed reducing strongly the energy consumption and enhancing the transport and information storage speed and capabilities, industry has to realize that the time for "beyond CMOS" [5] approach has arrived. In this case, we are talking about a new generation of devices not based on present technologies, mainly that do not rely on transistor implementation or more generally on charge transport and modulation. We think that the new emerging phenomena highlighted in 2D materials, what we call exotic properties, can help us to project ourselves in this new world where the labelling of devices and systems will not be the same and where transport and storage (of charges but mainly of spin information) will be achieved in parallel through a new generation of devices. This is the incredible value of the 2D materials and we have to move in this direction.

6.3 Some considerations

All the phenomena presented show interesting aspects related to the exotic physical properties involved. These properties give us new perspectives in the field of physics. In some cases, they need new mathematical approaches in order to describe them and they are quite far from our everyday perception of things. In other cases, such as for straintronics or black phosphorus (BP), we are not so far to implement the properties of materials in real devices. The different phenomena can be classified by also relating them to the potential realm of innovation. For example, in case of topological insulators (TI) we can imagine their implementation in new kinds of devices that will allow the final fabrication of an all-spin-related computer able to reduce dramatically the energy consumption increasing dramatically the calculation speed and the information storage. In this case, we are talking about a new generation of components that project us in the beyond CMOS area. However, for 2D TIs, maybe the nearest innovation in time is in the field of thermoelectricity and so in the domain of the more-than-Moore vision. It is quite difficult for researchers to understand that the intrinsic properties of 2D TI materials are by themselves largely sufficient to achieve results, which are largely beyond the present technologies ($zT \gg 3$). We would like to highlight an anecdotic thing in this context. We submitted a project in the framework of the Horizon 2020 European call last year. The project dealt with 2D TIs (specifically stanene, bismuthene and tellurene) and their potential implementation in thermoelectric generators. This project was, in great part, about the exploitation of stanene, which, as it has been shown in Chapter 1, owns intrinsic extraordinary features for thermoelectric, thanks to its TI character. This project did not get through and the main reason was related to the fact that the experts were not able to understand the methodology allowing to reach a zT larger than 3. Up to now, other teams mainly working on 3D materials or thin-film have worked on the creation of defects to reduce the thermal conductivity. In case of 2D TIs such as stanene and plumbene (i.e. Leadene), there is no specific methodology or engineering technology to use considering its intrinsic extraordinary properties. The experts, indeed, were not able to perceive the existence of a material with these characteristics and were not able to see beyond the existing technologies. This fact highlights the impact of the new physics on our perception of things and the fact that we have to open our mind in order to reach breakthrough innovation.

In case of the discovery of the magic angle, a new branch of physics has been created. The elegance of the physics behind the phenomenon has also its own importance in its success. This discovery has allowed the emergency of new theory on superconductivity and on the way to induce it. However, the race for publications is destroying the intrinsic significance of the discovery from a physical point of view. We would like to make an analogy using a completely different field. If we consider "Art Nouveau style" from the end of the nineteenth century to the first decade of the twentieth century, it allowed building incredible houses, mansions characterized by the beauty and elegance of the style inspired by great artist such as Victor Horta in Brussels. The problem was that the success automatically produced a large panel of works claiming themselves as "Art nouveau production" but without presenting any artistic benefit. These works were dubbed as "Noodle style" polluted the artistic production of Art Nouveau leading to its end. This is exactly what it is happening for the magic angle. A large panel of teams are publishing too many papers in this field pushed by the hot topic and by the potential impact of the reviews. The real issue is that a great part of these works have indeed a quite low added scientific value and innovation potential and make us lose the focus on the more interesting scientific aspects, as exactly what happened for "art nouveau style" and its artistic content, dispersed in a lot of insignificant productions. Moreover, the race to be the first to make the teams not analyze in a deeper way their discoveries. In case of the magic angle, a great part of the teams are working on three-layer or multilayer structures without having disclosed all the physics aspects behind what happens in bilayer graphene, the most simple systems. This does not allow understanding completely the potentialities of this topic.

In case of valleytronics, the potential to store three data (charge, spin and valley) in only one particle is real shift in the paradigm for the next generation of optically driven supercomputers. This could really open the doors to the quantic computer with completely new architectures. We are in this case in a completely new vision for the future of devices and in the realm of the beyond CMOS. However, this topic, at this moment, is not responding to the initial promises. The technical hurdles seem to be too important in terms of materials quality but also in terms of implementation of the phenomenon. Indeed, it seems too difficult to harness the information in a suitable way to be correctly stored because of the too low life-time of the particles states characterizing the phenomenon. An implementation is not impossible but we have to talk a very long-term research to think in terms of real devices. However, we have to recognize that the physics behind the valleytronics has opened a new field of research that even if it would not directly lead to applications, has allowed to do a step forwards in the knowledge of 2D materials, and more specifically of TMD 2D materials. We have to recognize that the implementation in crypto-technologies, thanks to the inherent properties of TMDs (the potential to recognize easily a specific polarization), could have a strong impact in the near future. However, we are again in the case of a "more-than-Moore" innovation which adds new functionalities without introducing a shift in the paradigm.

Maybe the materials that have potentially easier implementation in device at a quite short/medium term is BP. Its implementation in optoelectronic devices is straightforward and quite easy, simply considering its physic properties and the potential to achieve materials with ad hoc direct energy gap. In this case, we deal with more-than-Moore applications considering that we want to integrate the materials in already existing optoelectronic systems. Indeed, inversely what can happen for TIs, very probably BP will not give rise to a new generation of devices but it will be able to add new functionalities to the existing ones ("more than Moore"). To reach this objective, the main bottleneck concerns its "production" and its capacity to be handled correctly considering its strong reaction with oxygen and environmental humidity. Passivation technologies have been developed but this could be potentially at the origin of a large increase in the cost for final device fabrication or integration. This is a major drawback. Maybe the fabrication of composite where BP is inherently protected could be a suitable route to follow as demonstrated in case of solar cells.

Finally, straintronics is a technology that can open new field to change the properties of 2D materials in a quite deterministic way. This technique can be exploited in van der Waals structures to give rise to new specific features. It is not a new technique for semiconductor industry that has largely exploited heterostrain to achieve ad hoc electronic architectures. However, consider the specificity of 2D materials, we can think that using the strain we can, for example, allow new properties to emerge. We can imagine using strain to make emerge topological features in graphene, for example, at temperatures which are easier to reach (not only fraction of $1°K$). This could be a real shift in the paradigm and can be associated with a beyond CMOS vision. But we are very far from it. Indeed the main issue is the way in which the strain is induced in nanomaterials that constitute a real bottleneck. We can point out that magic angle measurements are strongly affected by strain that is not adequately mastered, as highlighted in Chapter 5. Therefore, the reproducibility of results does not allow us to think about a potential industrial implementation at a quite short term. It is clear that if we want to make real the Geim's "lego vision" for vdW structures composed of different materials with different lattices, we need to master perfectly this technique and this is not true at the moment.

Finally, there are some considerations about the tendency of using 2D materials to implement a "more-Moore" vision. The first attitude of a large number of scientific teams has been to employ 2D materials for a new generation of transistors. We think that this is really a loss of time. We do not need a new generation of transistors based on 2D materials for three main reasons. Firstly, performances are not better than the existing devices. Secondly, miniaturization of these devices is very difficult considering the limit of the present CMOS technology. The development of new materials need at less 5 years of updating of the process fabrication pilot lines. During this time, the present technology will have improved its performances making vain all the efforts. Thirdly, why to exploit new materials simply to make exactly the same devices without exploiting their intrinsic properties that can allow us to conceive a new generation of devices based on other physics phenomena and not simply on charge transport? Only in this way the real potential of 2D materials will be disclosed and a new generation of devices with different functions based on new physics will really able to change our way of seeing technology. We are talking, for example, spin-logic circuit that is able to store and transport information reducing dramatically the energy consumption. This will not be possible for all materials or phenomena developed in this book but we have to see beyond our everyday vision and to think out of the box. Not all innovations will go through the developments of new transistors!

6.4 Roadmaps

In the next sections, we will deal with the roadmaps for different applications of materials in specific applications. It is clear that considering the extremely rapid developments in specific fields, these roadmaps can evolve. However, thanks to that we are able to point out which are the turning points for each technology and when if the interest on the topic is at risk when confronted with technological or also cost issues on industrial implementation.

6.5 Topological insulators and thermoelectric

Firstly, we analyse the case of thermoelectric generators. The potential for high-efficient thermoelectric materials is very huge in case of 2D TIs. The main issue will concern the capacity to grow materials with the necessary quality especially in case of stanene. A major hurdle is also constituted by the test phase considering the difficulties in growth and translating the layers, and finally, their passivation. However, the potential is so important that we think that these issues will be solved and, even if at a long term, we will be able to have devices exploiting 2D TI features, for example, in Internet of things energetically autonomous devices that need extremely high miniaturization and high efficiency. Leadene/plumbene has even stronger capacity to highlight TI features at ambient temperature considering its Z number. However, in this case, we have to deal with considerations related to the toxicity of the materials and its psychological impact on public. Indeed, the quantity of materials will be extremely low compared to what it is presently used (which is also toxic) but the opposition to develop technologies based on lead could be a major hurdle. Now, the main challenge is related to the reproducible growth of these materials on adequate substrate preserving TI features. Theoretical studies combined with experiments could surely help to perform this effort, even if at the moment these kinds of scientific contributions are quite limited in number (see Fig. 6.2).

Another potential implementation is in electronic circuits based on spin. In this case, the protected edges of 2D TIs materials can act as a sort of highway for spin information which is robust against potential non-magnetic defects. This will give rise to a new generation of devices and systems with a huge reduction in energy consumption and with an enhanced speed that can lead us to the sacred Graal of the whole spin-based computer. However, there are some bottlenecks that, as for other materials such as graphene, have to be overcome. For example, the design of new devices with completely new features compared to existing ones, such as

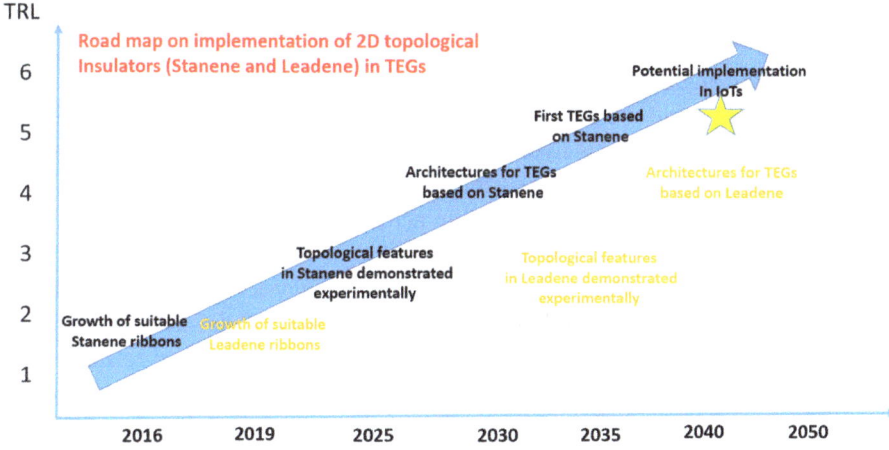

Fig. 6.2: Roadmap for implementation of 2D TI materials in TEGs.

common transistors, that can exploit the characteristics of 2D TI materials and target a new component based on spin logic. This will be in our opinion the turning point and this is very far from reality now. This is a very long-term vision that however could have a strong impact in our future. As previously highlighted, the main efforts have to be focused on the quality of the materials growth considering that the most interesting ones, such as stanene, are artificial materials (see Fig. 6.3).

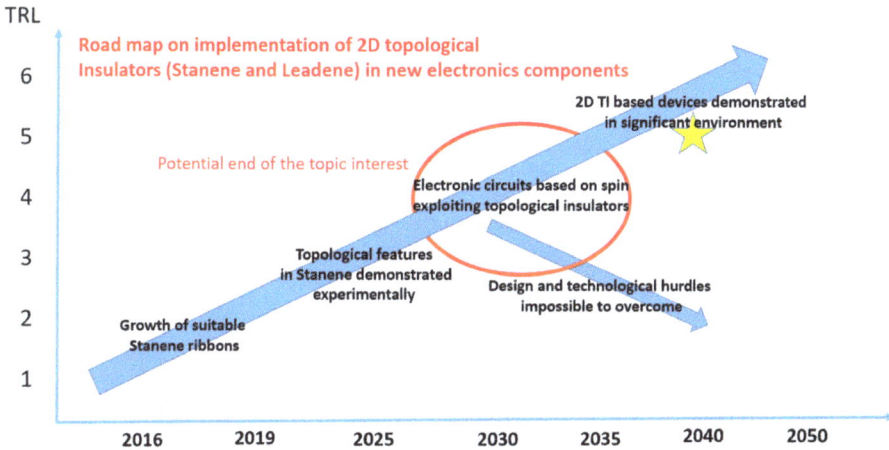

Fig. 6.3: Roadmap for implementation of 2D TI materials in new electronic components (spin-based logic).

6.6 Magic angle and superconductivity

In case of magic angle physics, we are really at the beginning of a new adventure. It is very difficult to imagine a real implementation of this phenomenon in devices. We can only try to formulate some hypothesis. It is clear that until the effects will not be completely harnessed, and it is impossible to make a realistic roadmap. The problem is also related to the temperature needed to achieve superconductivity, which is too low to think to fabricate suitable devices exploiting this phenomenon. We can say that the convergence of technical interpretation of the physics and capacity to identify potential application issues do not allow us to be very optimistic. Much work is needed and more studies on the physics of the simplest systems could be necessary. It is clear that the possibility to achieve superconducting materials that can be exploited in supercomputers, or generally in a new generation of electronic circuits, can lead to open new fields of innovation and a revolution in the field (see Fig. 6.4).

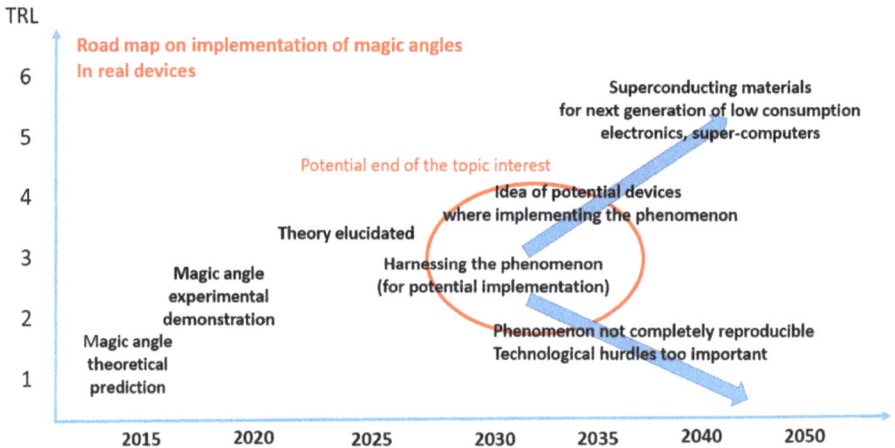

Fig. 6.4: Roadmap for implementation of "magic angle materials" in real devices.

6.7 Valleytronics and information storage

In case of valleytronics, the potential implementation in new ways to achieve information storage discloses an incredible potential of innovation. The possibility to store three data in one particle is a real shift in the paradigm and can lead to the realization of new supercomputing systems based on optical signals. However, the implementation of this phenomenon is very difficult considering the strong limits related to the lifetime of the states of the particles after optic excitation. This does not allow now to think about suitable devices exploiting this phenomenon. The hope is to try to find new ways of harnessing the information based on valleys that present technologies that do not permit to imagine. This will be the key factor that will push for development in the field or will stop definitively the interest in the topic. We think that it will be a long-term work that, however, has to deal with the loss of the hype of the topic leading to a reduction of research funding on it (see Fig. 6.5).

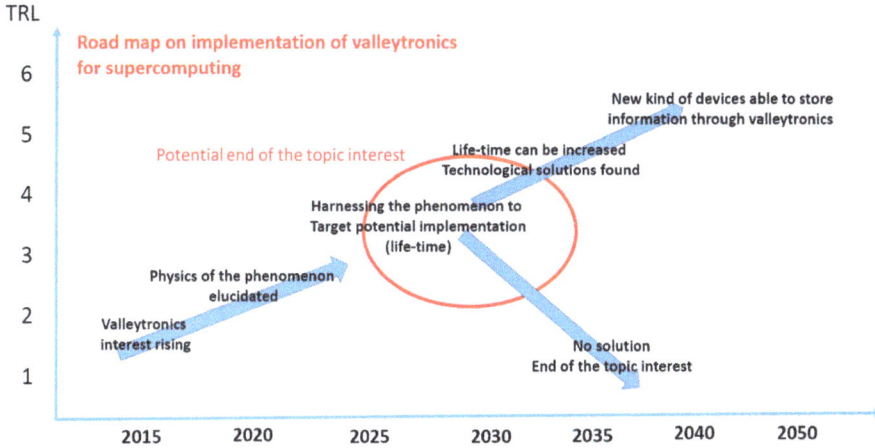

Fig. 6.5: Roadmap for implementation of valleytronics in 2D materials in real devices.

A good point in favour of implementation of valleytronics in devices is their utilization in crypto-technologies. As already stated in Chapter 3, quantum key distribution (QKD) is a logical application considering that in this case information is usually encoded on single photons. For this specific application, it is necessary to create/detect photon with a specific polarization. Valleytronics is implemented through the intrinsic properties of TMDs as in MoS_2, which can be easily exploited with the possibility to achieve miniaturized optical systems at microchip level avoiding the integration of encumbering optical elements. The QKD, thanks to this potential miniaturization brought by the utilization of TMDs, can be integrated to secure communication in mobile phone, tablets and also the information about personal health care. This could be a quite short-term horizon for the final utilization of valleytronics in real devices.

6.8 Black phosphorus and optoelectronics

BP has immediately been identified as a material that opens important possibilities in the field of optoelectronics. Indeed, BP, as already highlighted, has the very peculiar properties to have band gap changing as a function of its thickness. However, the direct gap is always at the same place of the band diagram, in opposite of what happens, for example, for TMDCs (where also it is not always direct). It is clear that the great

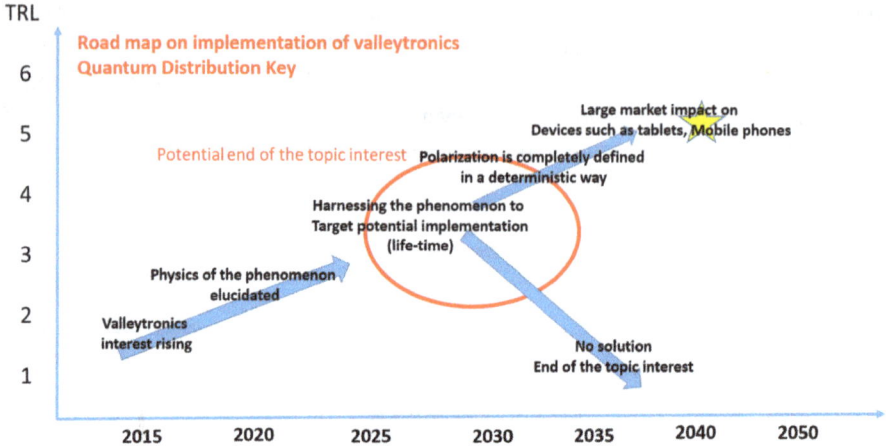

Fig. 6.6: Roadmap for implementation of valleytronics in crypto-technologies.

bottleneck concerns the capacity to handle the material correctly. This is the only factor that can slow the research efforts. If research will get through this turning point, the implementation in devices will be straightforward and the real applications will be at a quite short term (see Fig. 6.6). Indeed, we stay in realm of the "more-than-Moore" approach. A quite immediate application could be in ultra-miniaturized gas sensors with absorption spectra in the infrared region. However, also the application in the context of high-energy micro-supercapacitors seems to be very promising considering the first very promising results, especially for wearable devices. This is one of the main characteristics of 2D materials. Their potential to be deposited on flexible/conformable substrates and to be robust against compression and torsion. This is a typical application where BP can add a new functionality. The last but not least is the utilization of BP for spintronics. Thanks to the very high mobility of BP, we can imagine that the implementation in spintronics based devices can be a long-term view. However, the processability and passivation do not have to influence layer features. Some teams are working actively on this material and maybe we can imagine real devices based on spintronics based on BP, even if not at a short term. We can conclude saying that maybe the most interesting applications concerning the fabrication of composites using BP or more precisely phosphorene. In this case, the material is inherently protected inside the composite. This is true in case of fabrication of solar cells or also in batteries.

6.9 Straintronics and different potential implementations

Straintronics cannot be defined as a technology that will implement a Beyond CMOS vision. As previously pointed out, one of the main interests could be its utilization in vdW structures composed by layers of different origin. Thanks to that,

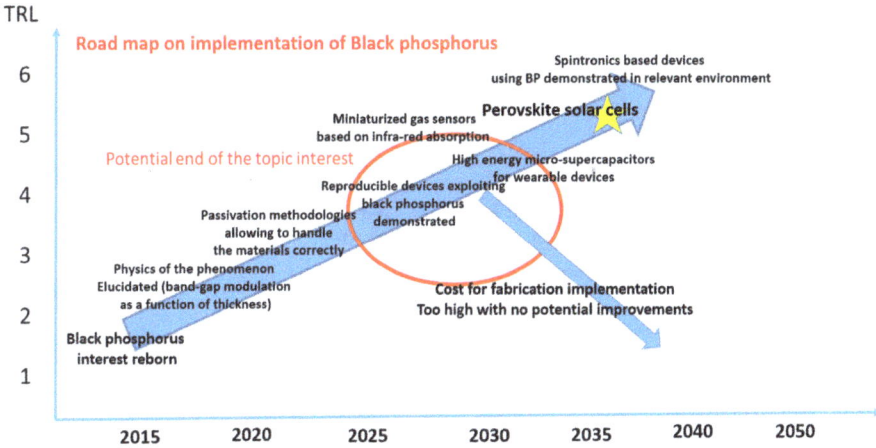

Fig. 6.7: Roadmap for implementation of black phosphorous in optoelectronic devices.

we can implement all the different kinds of strains such as in-plane but also het-ero-strain, already commonly used by semiconductor industry to achieve architec-tures with ad-hoc properties. However, there are some technical bottlenecks (not small ones) to solve. Firstly, the fabrication of vdW structures is not optimized yet. Secondly, applying a reproducible strain to a 2D material is not an easy task with problems in terms of reproducibility of results. For these reasons, we think that the implementation of straintronics in vdW structures is also a long-term re-search topic that will meet its turning point when the scientists will be able (or not) to fabricate reproducible vdW structures with reproducible functionalities (see Fig. 6.7). This vision is clearly a more-than-Moore vision, where we do not conceive new concepts of devices but we implement new function to existing ones. However, a potential beyond CMOS vision could be adopted if, in the future, researchers will be able to induce topological features in 2D materials simply, for example, applying in-plane strain. No team is working on this idea but we think that maybe this could also be a sort of new revolution for this approach.

6.10 Classification of the potential innovations

We would like to end this section summarizing all the applications analysed and their potential of innovation. We have the domain in which all of the innovation can be classified as a function of the Moore law (see Fig. 6.8). This analysis is not exhaustive but points out only some of the major fields of applications taken into consideration. In red, we have highlighted the potential breakthorugh that can lead potentially to a shift in the paradigm.

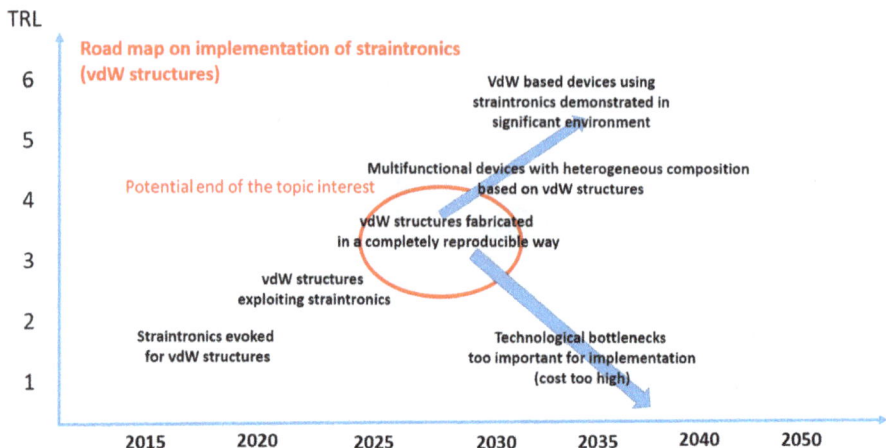

Fig. 6.8: Roadmap for implementation of straintronics in vdW structures.

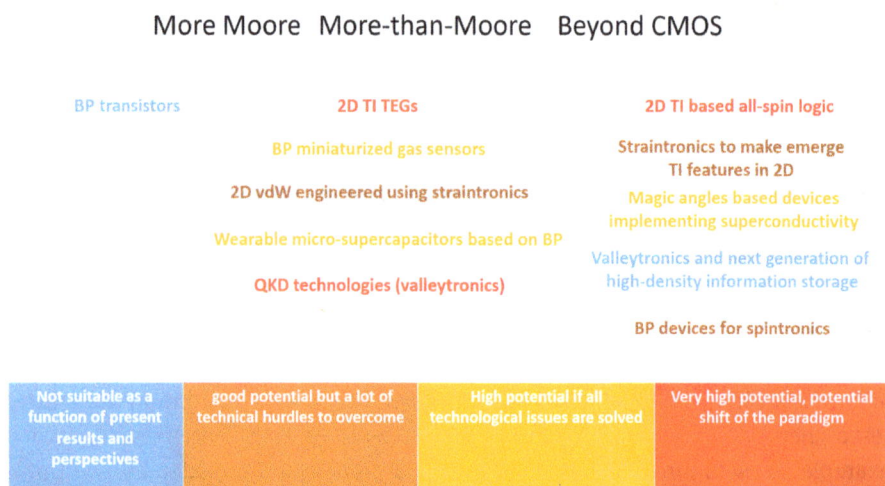

Fig. 6.9: Innovation classification.

We can observe that a lot of innovations dealt with the domain of beyond CMOS and this is really a good news (see Fig. 6.9). The only problem concerns the fact that all these topics need to be deeply studied and not simply used as an object for immediate publication in high-rate papers. We need a larger collaboration between theoretical scientists, experimentalists and also companies in order to understand, even if at long time, which are the main bottleneck and identify the coherent route in research.

References

[1]   Intel Unleashes Next-Gen Enthusiast Desktop PC Platform at Gamescom. Intel Newsroom. Retrieved August 5, 2015
[2]   law Moore, Gordon E. (1965-04-19). "Cramming more components onto integrated circuits". Electronics. Retrieved 2016- 07-01.
[3]   www.itrs2.net/uploads/4/9/7/7/49775221/irc-itrs-mtm-v2_3.pdf
[4]   http://www.itrs2.net/uploads/4/9/7/7/49775221/irc-itrs-mtm-v2_3.pdf
[5]   https://irds.ieee.org/images/files/pdf/2020/2020IRDS_BC.pdf

Index

https://doi.org/10.1515/9783110656336-008